ADVANCES IN ENERGETIC DINITRAMIDES

An Emerging Class of Inorganic Oxidizers

T0350219

ADVANCES IN ENERGETIC DINITRAMIDES

An Emerging Class of Inorganic Oxidizers

Ang How Ghee
G Santhosh

Energetic Materials Research Centre
Nanyang Technological University

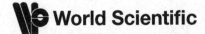
World Scientific

NEW JERSEY · LONDON · SINGAPORE · BEIJING · SHANGHAI · HONG KONG · TAIPEI · CHENNAI

Published by

World Scientific Publishing Co. Pte. Ltd.

5 Toh Tuck Link, Singapore 596224

USA office: 27 Warren Street, Suite 401-402, Hackensack, NJ 07601

UK office: 57 Shelton Street, Covent Garden, London WC2H 9HE

British Library Cataloguing-in-Publication Data
A catalogue record for this book is available from the British Library.

ADVANCES IN ENERGETIC DINITRAMIDES
An Emerging Class of Inorganic Oxidizers

ISBN-13 978-981-277-203-9
ISBN-10 981-277-203-0

Typeset by Stallion Press
Email: enquiries@stallionpress.com

Printed in Singapore.

PREFACE

Energetic materials are key ingredients in propellants and explosive formulations. They are of strategic importance to modern defense system and weaponry. The rapid growth and increased demand for advanced energetics have yielded new classes of energetic materials. Among them is the dinitramide salts which belong to an emerging class of inorganic oxidizers whose unique dinitramide anion possesses both resonance-stabilized structures as well as high oxidizing power. Among the many dinitramide salts that have been reported, the outstanding ones that have attracted extensive interests are ammonium dinitramide (ADN) and guanylurea dinitramide (GUDN).

ADN is emerging as an outstanding candidate to replace the conventional AP as a rocket propellant oxidizer. It offers enhanced performance in terms of specific impulse (I_{sp}). Its improved lift capacity would enable the rockets to carry larger payloads. It is also environmentally friendly and does not release toxic gaseous products during combustion, in contrast to the damaging effects of chlorine and hydrogen chloride produced by AP on the earth's ozone layer and green vegetation. The excess oxygen content in the dinitramide moiety can be used in solid propellants as well as explosive formulations to effectively oxidize the fuel elements such as binders, metals and metal hydrides. ADN is totally smoke-free upon burning, and its plume is not easily detected by radar.

GUDN on the other hand, is an insensitive thermally stable material which has attracted much attention for possible application as gas-generators in automobile airbags. In combination with oxidizers such as FOX-7 and RDX mixed with a binder system, it also finds application as insensitive gun propellants.

Notwithstanding the advantages of ADN as a promising future propellant oxidizer, much more research is needed in order to shape ADN as a green and promising candidate with enhanced performance in place of AP. Among the problems that need to be solved are: its relatively low temperature of decomposition and thermal stability in processing; crystal modification of ADN to attain a theoretical maximum density; hygroscopicity under high humidity; incompatibility with certain isocyanates, energetic binders and additives; less than ideal burn rate and combustion instability; cleaner and cheaper methods of synthesis.

Over 400 research papers related to ADN and other dinitramide salts have been published so far. A few reviews dedicated to the synthesis, combustion and thermal decomposition of dinitramide salts have appeared in the literature. The lack of a comprehensive reference on the chemistry and technology of ADN and other dinitramide salts has prompted us to write this monograph.

Ang How Ghee
G Santhosh
Singapore, August 2007

CONTENTS

Chapter 3   Synthesis of Other Dinitramide Salts   25

Chapter 4   Characterisation of ADN   35

Chapter 1

INTRODUCTION

Although patents and publications on ammonium dinitramide (ADN) started to emerge since the late 1980s, it is now known that the outstanding inorganic oxidizer was first discovered in the former Soviet Union some ten years earlier.

ADN and many more dinitramide salts that have been reported contain the unique dinitramide anion, whose structure is illustrated in Fig. 1.1.

ADN has attracted wide interest because of its potential as an oxidizer in solid rocket propellant application.[1,2] It is a clean and environmentally benign solid oxidizer envisaged as a replacement for the conventional oxidizers viz. ammonium perchlorate (AP) and ammonium nitrate (AN) used in composite solid rocket propellants.[3]

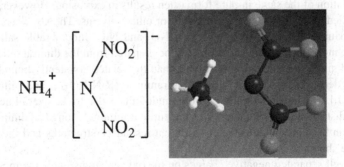

Fig. 1.1. Structure of ADN.

After ADN, many other dinitramide salts have also been synthesized containing different cations ranging from organic to inorganic and metal complexes. The most notable among them is the guanylurea salt of dinitramide, which shows unique promise as a possible candidate for use in automotive air bags. It displays a range of desirable properties that would make it competitive on the ground of safety and environmental regulations.

1.1. Physical and Chemical Properties of ADN

ADN is a pale yellow to off-white crystalline hygroscopic solid. The magnitude of hygroscopicity is higher than that of AN. To prevent ADN from absorbing moisture, it has to be handled below a relative humidity of 55%. It is highly soluble in polar solvents such as water, methanol, acetone, acetonitrile, isopropanol etc.; sparingly soluble in ethyl acetate, methyl isobutyl ketone, n-butanol etc.; and insoluble in ether, methylene chloride, benzene, chloroform etc. From solvents, it crystallises as thin prisms, needles or platelet shaped crystals. It shows no phase transitions at atmospheric pressures and a high pressure reversible phase transition is observed at 2 GPa ± 0.2 GPa.[4] The new high pressure phase β-ADN is monoclinic. The physicochemical properties of ADN are summarised in Table 1.1.

ADN has a melting point in the range of 92°C–95°C followed by decomposition in the temperature range of 130°C–220°C. It forms an eutectic with AN, the melting point of the ADN/AN (70/30 mol %) eutectic is 55°C.[4] ADN-water forms an eutectic at 58% of ADN with a melting point of −15.3°C.[11]

1.2. Reactivity and Different Forms of Dinitramide

The acid form of ADN i.e. dinitramidic acid $[HN(NO_2)_2]$ is a very strong mineral acid with pKa = −5.62. It is a colourless unstable liquid under ambient conditions and isolation of the same in pure form often results in explosion. However, it can be handled safely in dilute form in water or other solvents. The key decomposition products of dinitramidic acid are HNO_3 and N_2O. Innumerable salts with organic, inorganic and metal cations can be prepared from the dinitramidic acid.

The dinitramide salts are more stable than the related covalently bound N,N-dinitro derivatives such as alkyl dinitramines $[R-N(NO_2)_2]$ and nitramide $[NH_2NO_2]$. The higher stability of dinitramide salts is due to the overall negative charge distributed by resonance over the entire dinitramide ion. The dinitramide anion can be represented by many different resonance structures and the stable ones are shown in Fig. 1.2.

The well separated negative charges on the oxygen atoms make the most substantial contribution to the total electronic structure, which agrees with the calculated

Table 1.1. Physicochemical Properties of Ammonium Dinitramide.

Property		Reference
Appearance	Pale yellow or off-white crystals	–
Melting point	92°C–95°C	[3]
	55°C (70/30 ADN/AN)	[4]
Density	1.80 g/cm^3–1.84 g/cm^3	[5]
Decomposition temperature	127°C	[5]
Temperature of ignition	142°C	[6]
Enthalpy of formation	−150.6 kJ/mol	[5]
Dissociation energy	255 kJ/mol ± 12.5 kJ/mol	[5]
Friction sensitivity	72 N	[5]
Impact sensitivity	5 Nm	[5]
Electrostatic discharge	0.45 J	[6]
Figure of insensitivity (gas evolved, mL)	30(2.3)	[6]
Vacuum stability (80°C/40 hour, ml/5g)	0.73	[6]
Yield of gas, moles/100 g	4.03	[7]
UV absorption maxima in water	212 nm and 284 nm	[3]
pKa (acid form)	−5.62	[8]
Heat of combustion	980 kJ/mol	[9]
Transport classification	1.1D	[10]
Water solubility	78.1% at 20°C	[11]
Particle shape	Needles, prisms or	
	platelets (crystallised)	[12]
	spheres (prilled)	[12]
Oxygen balance	+25.79%	[11]
Oxygen content	51.6%	–
Nitrogen content	45.16%	–

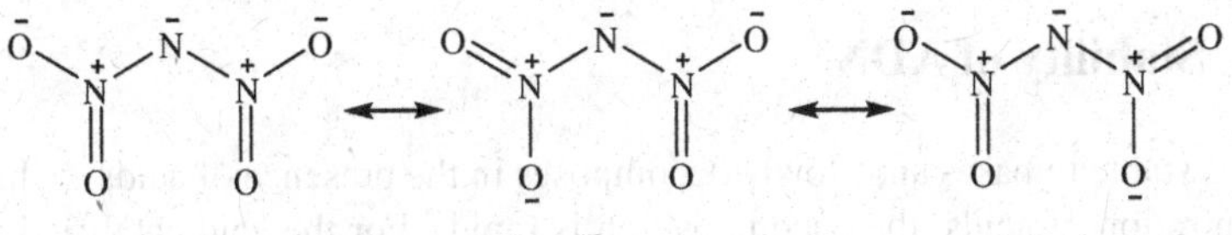

Fig. 1.2. Resonance structures of the dinitramide anion.

atomic charges (RHF/6-31*6). *Ab initio* calculations at the MP2 level using 6-31G** basis set for the anion showed that the minimum energy structure has a C$_2$ symmetry in which the oxygen atoms are above and below the plane of the three nitrogen atoms.

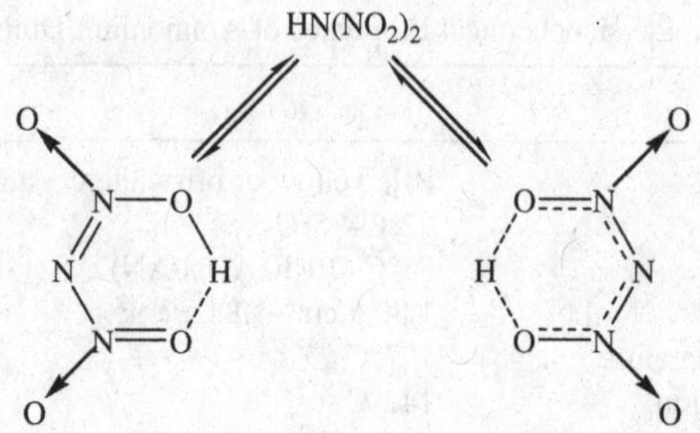

Fig. 1.3. Covalent forms of dinitramidic acid.

Depending on the method of preparation, the dinitramide anion can exist in three different forms.[13] The ionic form which exists in aqueous solutions can be easily detected by the UV absorption bands at 225 nm and 285 nm and a low intensity shoulder at 335 nm. The ionic form exists in polar solvents including the solutions of strong acids. The second and third covalent form of dinitramide can be detected in weakly polar organic solvents such as ether, benzene, chloroform, dichloromethane, dichloroethane, or dioxane. The second covalent form is detected when potassium dinitramide in ether is saturated with HCl and concentrated under vacuum. The dinitramide prepared by this method has the structure of the N-H form in equilibrium with the aci-form having an intramolecular hydrogen bond. It has an absorption maximum at 223 nm. The third covalent form is detected at 243 nm–250 nm in the UV spectrum when the potassium dinitramide is dissolved in 65% H_2SO_4 and extracted with dichloroethane. The structure of this dinitramide is considered as an aci-form with strong intramolecular hydrogen bond between the oxygen atoms.[13] The two covalent forms of dinitramidic acid are shown in Fig. 1.3.

1.3. Stability of ADN

ADN is stable to bases and slowly decomposes in the presence of acids. In higher concentration of acids, the decomposition is rapid.[3] For the acid catalyzed reaction, the rate is directly proportional to the acidity. In acid medium, the decomposition occurs via protonation of the dinitramide anion. The presence of moisture and ammonium nitrate strongly affects the stability of ADN, as samples containing moisture and AN are less thermally stable than the pure ADN. The thermal stability of molten ADN is low when compared to solid ADN. The decomposition of ADN is autocatalytic above its decomposition temperature. The decomposition products of ADN viz. HNO_3, N_2O, NO_2 etc. catalyze the decomposition. Anomalous decomposition of ADN is observed at 60°C, where the rate of decomposition exceeds several folds than the rate of decomposition at 80°C.[14] ADN is incompatible towards certain isocyanates used in the curing of propellants. However, the cured propellants based on ADN doesn't show any stability or compatibility issues. Experiments with 150 mM of aqueous solutions of ADN at pH 4, 7 and 9 showed no decomposition over a period of 37 days in the dark at 60°C.[15] The dinitramide anion in aqueous solutions are indefinitely stable at 25°C in the dark under abiotic conditions.

1.4. Light Sensitivity of ADN

ADN is sensitive towards light and decomposition is observed for the samples exposed to light.[16,17] ADN absorbs strongly in the UVA region (320 nm–365 nm) of the spectrum and solutions of ADN photolyzes in sunlight to give nitrate (NO_3^-), nitrite (NO_2^-) and nitrous oxide (N_2O). The pH of the ADN solution greatly affects the stability under photolysis. At pH 11, the ratio of NO_2^-/NO_3^- is about 20 and at pH 2, the ratio declined to 0.5. Solid samples of ADN must always be stored in dark colored bottles to prevent the decomposition in the presence of light. 10 μM–50 μM solutions of ADN had a half life of about 3 min–4 min in sunlight. A first-order rate of NO_2 formation was observed[18] during the UV-photolysis of ADN with a rate constant value of 0.013×10^{-7} min^{-1}. In summer noon light at 40°N latitude, the theoretical minimum half life for ADN is about 20 sec.[15] Photolysis of 100 μM ADN solutions in methanol, acetone, DMF and acetonitrile produced equal amounts of NO_2^- and NO_3^-. The rate of reaction is fast in acetone because it absorbs all UV radiation below 335 nm and transfers the energy to ADN via triplet-singlet interaction. Direct photoreactions are observed for the other solvents. The aqueous decomposition pathways for ADN in the presence of light are shown in Scheme 1.1.

The photolysis of solid ADN produced NO_3^-, NO, N_2O and N_2 with small amounts of NO_2^-. N_2 has been formed as a result of a redox interaction between NH_4^+ and NO_2^-.

1.5. Sonochemistry of ADN

Qadir *et al.* have shown that solutions of ADN decomposes when subjected to sonication under varying pH and in presence of Al.[19] No decomposition is observed when the neutral ADN solution is sonicated. However, sonicating a

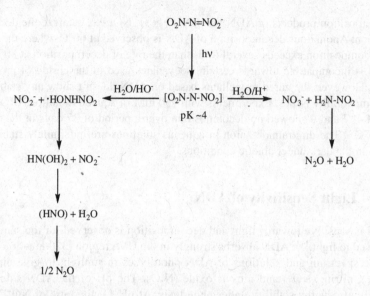

Scheme 1.1. Pathways for the photolysis of aqueous dinitramide ion.

solution of 151 µM ADN/10 mM NaOH and 5 mg/ml of Al powder for 60 min showed that nearly 86% of ADN was decomposed, while reducing the concentration of ADN to 50 µM, 100% decomposition was achieved in 20 min. The authors propose a decomposition mechanism for ADN under the sonication conditions. They suggest that the ultrasound technique may be adopted for the safe disposal of aqueous solutions of ADN.

1.6. Radiolysis of Dinitramide

Radiolysis of aqueous solutions of potassium dinitramide was studied by Milekhin *et al.*[20] The decomposition of dinitramide was followed by monitoring the absorbed dose of radiation by chemical and spectral methods. The effect of addition of alcohols and the saturation of the solutions with N_2O was also studied. The irradiation results show that the conversion is 0.93 ± 0.02 dinitramide ions per 100 eV of energy, and the introduction of 0.1 M ethyl or methyl tert-butyl alcohol increases the conversion to 4 ± 0.2 and 1.56 ± 0.09 ions respectively per 100 eV of energy. However, the saturation of the solutions with N_2O decreases the conversion to 0.3 ± 0.02 ions/100 eV of energy. Their results indicate that the water radiolysis products reduce the dinitramide anion to nitrite ions, where the conversion or the chemical yield depends on the amount of radiation absorbed.

1.7. Biotransformation of ADN

The aerobic and anaerobic biotransformation of ADN in a source of microbial population viz. in a water/sediment mixture showed that no loss of ADN was observed after 60 days of incubation.[15] The results indicate that either ADN is toxic to the aquatic organisms or they have no need to use ADN as a nitrogen source. Investigations of different concentration of ADN in the presence of a bacterial growth medium viz. Tripticase soy medium (TSM) showed that it followed pseudo first-order kinetics. At high concentration of ADN with 500 ppm TSM, ADN loss was observed over a 16-day period and no loss was observed in less concentrated ADN solutions. The results show that ADN is not toxic to aquatic organisms.

1.8. Toxicity of ADN

The toxic properties of ADN has been tested on animals such as rats, mice, rabbits etc. *In vitro* toxicity of ADN has been studied by Dean *et al.*[21] Enzyme leakage assay test indicates that 2.7 mM of ADN concentration damages the cell membrane. It was also observed that cells exposed to ADN have the potential for directly affecting cellular DNA. The radical effects of ADN on human liver slices were studied by Steel-Goodwin *et al.*[22] Acute injury to the liver cells were observed when these were in contact with ADN. Cells exposed to ADN produced free radicals in presence of phenyl tert-butyl nitrone.[23] Incubation of hepatocytes with 2.8 mM of ADN for 24 hours is toxic to 50% of the cells. Toxicity/reproductive screens of ADN on rats by administering different doses of ADN in the drinking water showed a decrease in fertility on the pregnant rats.[24] ADN has an adverse effect on the development of embryos in rats.[25] At a concentration of 1mM or higher either slowed or arrested the development of embryos. Acute oral and dermal toxicity of ADN were performed by Kinkead *et al.*[26] ADN is readily absorbed by the skin resulting in numbness. The acute oral toxicity of ADN to rats is measured as $LD_{50} = 823$ mg/kg. The dermal toxicity tests on rabbits showed no adverse effects on the skin. Genetic toxicology effects on ADN salmonella/microsome mutagenesis, mouse lymphoma cell mutagenesis, *in vivo* mouse bone marrow micronuclei tests demonstrate that it is mutagenic to bacteria and mammalian cells and causes chromosomal damage.[27] The high water solubility of ADN poses additional risk to surface and groundwater contamination.

References

1. S Borman, Advanced energetic materials emerge for military and space applications, *Chem & Eng News*, 17 January, 18–22, 1994.

2. JC Bottaro, Recent advances in explosives and solid propellants, *Chem & Ind*, 249–252, 1996.

3. JC Bottaro, PE Penwell, RJ Schmitt, 1,1,3,3-tetraoxo-1,2,3-triazapropene anion, a new oxy anion of nitrogen: the dinitramide anion and its salts, *J Am Chem Soc* **119:** 9405–9410, 1997.

4. TP Russell, GJ Piermarini, S Block, *et al.*, Pressure, temperature reaction phase diagram for ammonium dinitramide, *J Phys Chem* **100:** 3248–3251, 1996.

5. U Teipel, T Heintz, HH Krause, Crystallisation of spherical ammonium dinitramide (ADN) particles, *Prop Expl Pyro* **25**(2): 81–85, 2000.

6. MD Cliff, DP Edwards, MW Smith, Alkali metal dinitramides. Properties, thermal behaviour and decomposition products, *29th Int Annu Conf ICT* **24:** 1–12, 1998.

7. ML Chan, R Reed Jr, DA Ciaramitaro, Advances in solid propellant formulations, in Solid propellant chemistry, combustion and motor interior ballistics, V Yang, TB Brill, WZ Ren (eds.), *Prog Astro Aero*, AIAA **185:** 185–206, 2000.

8. OA Lukyanov, VA Tartakovsky, Synthesis and characterisation of dinitramidic acid and its salts, in V Yang, TB Brill, WZ Ren (eds.), Solid propellant chemistry, combustion and motor interior ballistics, *Prog Astro Aero*, AIAA **185:** 207–220, 2000.

9. H Ostmark, U Bemm, A Langlet, *et al.*, The properties of ammonium dinitramide (ADN): Part 1, Basic properties and spectroscopic data, *J Energ Mater* **18:** 123–138, 2000.

10. P Sjoberg, R Wardle, T Highsmith, A cooperative effort to develop manufacturing processes for spherical ADN, 2001 Insensitive Munitions and Energetic Materials Technology Symposium, 466–470, 2001.

11. N Wingborg, Ammonium dinitramide — water: Interaction and properties, *J Chem Eng Data* **51:** 1582–1586, 2006.

12. LF Cannizzo, TK Highsmith, RB Wardle, *et al.*, Utilisation of ammonium dinitramide (ADN) in propellant formulations, *Technical Report No. A660504*, Thiokol Corporation, 1998.

13. VA Shlyapochnikov, NO Cherskaya, OA Lukyanov, *et al.*, Dinitramide and its salts 4. Molecular structure of dinitramide, *Russ Chem Bull* **43**(9): 1522–1525, 1994.

14. AN Pavlov, VN Grebennikov, LD Nazina, *et al.*, Thermal decomposition of ammonium dinitramide and mechanism of anomalous decay of dinitramide salts, *Russ Chem Bull* **48**(1): 50–54, 1999.

15. T Mill, R Spanggord, Fate assessment of new air force chemicals, *Technical Report No. AFOSRTR-97–06*, SRI International, 1997.

16. MK Beretvas, JP Hassett, SE Burns, *et al.*, Modeling the photolysis of ammonium dinitramide in natural waters, *Environ Toxicol Chem* **19**(11): 2661–2665, 2000.

17. BV Gidaspov, IV Tselinskii, MB Shcherbinin, Photolysis of dinitramide salts in solutions, *Russ J Gen Chem* **67**(6): 911–914, 1997.

18. MD Pace, Nitrogen radicals from thermal and photochemical decomposition of ammonium perchlorate, ammonium dinitramide and cyclic nitramines, *Mat Res Soc Symp Proc* **296**: 53–60, 1993.

19. LR Qadir, EJ Osburn-Atkinson, KE Swider-Lyons, *et al.*, Sonochemically induced decomposition of energetic materials in aqueous media. *Chemosphere* **50**(8): 1107–1114, 2003.

20. Yu M Milekhin, DN Sadovnichii, Radiolysis of an aqueous solution of potassium dinitramide, *Dokl Chem* **412**(2): 46–48, 2007.

21. KW Dean, SR Channel, *In Vitro* effects of ammonium dinitramide, *Technical Report No. AL/OE-TR-1996–0059*, 1995.

22. L Steel-Goodwin, KW Dean, DM Pace, *et al.*, Effects of ammonium dinitramide in human liver slices: An EPR/ENDOR trapping study, *Technical Report No. AL/OE-TR-1995–0161*, 1995.

23. ER Kinkead, RE Wolfe, ML Feldmann, Dose (and time dependent) blockade of pregnancy in Sprague-Dawley rats administered ammonium dinitramide in drinking water, *Technical Report No. AL/OE-TR-1995–0181*, 1995.

24. S Berty, L Steel-Goodwin, K Dean, *et al.*, The biological effects of ADN on hepatocytes: An EPR study, *Technical Report No. AL/OE-TR-1995–0173*, 1995.

25. LJ Graeter, RE Wolfe, ER Kinkead, *et al.*, Effects of ammonium dinitramide on preimplantation embryos in Sprague-Dawley rats and B6C3F1 mice, *Technical Report No. AL/OE-TR-1996–0171*, 1996.

26. ER Kinkead, SA Salins, RE Wolfe, Acute and subacute toxicity evaluation of ammonium dinitramide, *Technical Report No. AL/OE-TR-1994–0071*, 1994.

27. S Zhu, E Korytynski, S Sharma, Genotoxicity assays of ammonium dinitramide (1) salmonella/microsome mutagenesis (2) mouse lymphoma cell mutagenesis (3) *in vivo* mouse bone marrow micronuclei test, *Technical Report No. AL/OE-TR-1994–0148*, 1994.

Chapter 2

SYNTHESIS OF AMMONIUM DINITRAMIDE (ADN)

Owing to the prime importance of ADN as an oxidizer in solid propellants, its synthesis has received a great amount of interest. There have been several new and innovative synthetic routes developed by researchers worldwide. The following sections describe the methods of synthesis of ADN reported in the literature.

2.1. Synthesis of ADN from Nitramide

ADN was prepared from nitramide using NO_2BF_4[1a] or nitryl salts[1b]. The reaction was carried out at −20°C in anhydrous CH_3CN to form the dinitramidic acid which was later neutralised with ammonia to give ADN. The reaction is shown in Scheme 2.1.

The yield of ADN was about 60% when NO_2BF_4 was used. NO_2SO_3F and NO_2BF_4 gave excellent yield when CH_3CN was used as a medium. The solvents CH_2Cl_2 or EtOAc gave only poor yield and no dinitramidic acid was formed when hexane was used as a solvent.[1b] A solvent extraction procedure for the separation of ADN was given by Malesa et al.[2] A detailed analysis of the solid products during each extraction was performed and the results were summarised in their paper.

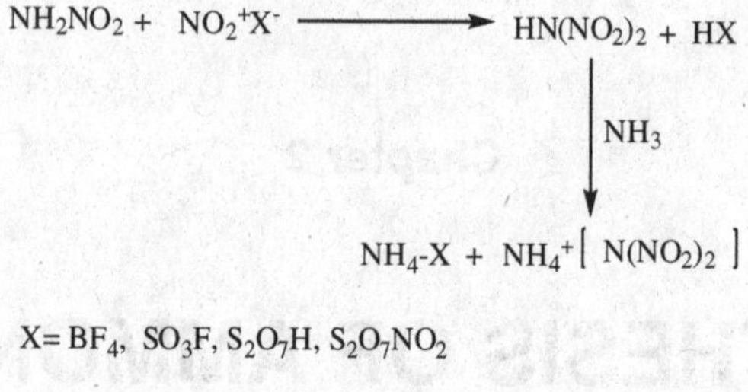

$$NH_2NO_2 + NO_2{}^+X^- \longrightarrow HN(NO_2)_2 + HX$$

$$\downarrow NH_3$$

$$NH_4\text{-}X + NH_4{}^+\left[N(NO_2)_2 \right]^-$$

$$X = BF_4,\ SO_3F,\ S_2O_7H,\ S_2O_7NO_2$$

Scheme 2.1. Synthesis of ADN from nitramide.

2.2. Synthesis of ADN from Ammonium Carbamate

The reaction of ammonium carbamate with nitronium tetrafluoborate was disclosed by Bottaro *et al.*[3] The reaction was carried out in the temperature range of −30°C to −40°C in CH_3CN to form the dinitramidic acid which was later neutralised with ammonia to give ADN as shown in Scheme 2.2. The yield of reaction was 15%.

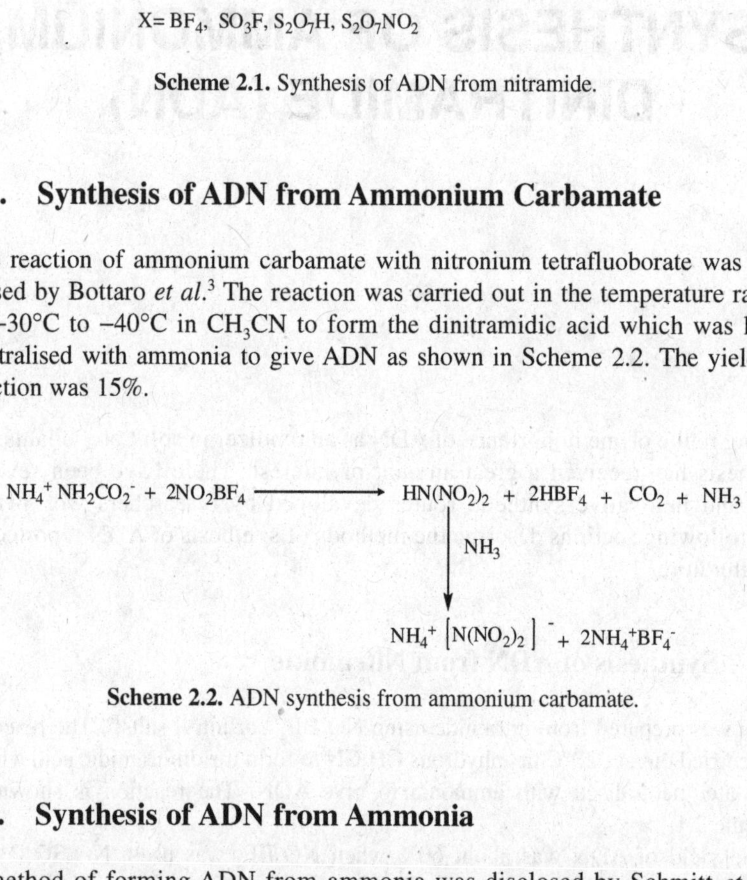

$$NH_4{}^+\,NH_2CO_2{}^- + 2NO_2BF_4 \longrightarrow HN(NO_2)_2 + 2HBF_4 + CO_2 + NH_3$$

$$\downarrow NH_3$$

$$NH_4{}^+\left[N(NO_2)_2 \right]^- + 2NH_4{}^+BF_4{}^-$$

Scheme 2.2. ADN synthesis from ammonium carbamate.

2.3. Synthesis of ADN from Ammonia

A method of forming ADN from ammonia was disclosed by Schmitt *et al.*[4] The nitration of ammonia in presence of either NO_2BF_4, N_2O_5 or $NO_2HS_2O_7$ in methylene chloride at −78°C forms ADN. The steps involved in the preparation are shown in Scheme 2.3.

The reaction is believed to proceed via the intermediate nitramide, which upon further reaction with the nitrating agent forms dinitramidic acid which on neutralisation with ammonia gives ADN. The yield of the reaction varies from 5%–20% depending on the nitrating agent used.

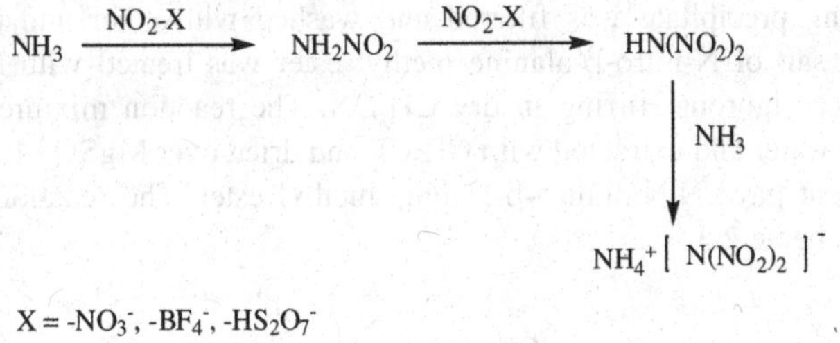

$$NH_3 \xrightarrow{NO_2\text{-}X} NH_2NO_2 \xrightarrow{NO_2\text{-}X} HN(NO_2)_2$$

$$\downarrow NH_3$$

$$NH_4{}^+[\ N(NO_2)_2\ ]^-$$

X = -NO$_3^-$, -BF$_4^-$, -HS$_2$O$_7^-$

Scheme 2.3. Synthesis of ADN from ammonia.

2.4. Synthesis of ADN from N,N-dinitro-β-aminopropionitrile

The synthesis of ADN was described by Lukyanov *et al.* starting from N,N-dinitro-β-aminopropionitrile.[5] To a solution of the same in dry dioxane, dry NH$_3$ was passed for 30 min at 8°C–10°C. The precipitated ADN was removed by filtration and washed with dioxane-ethyl acetate mixture and then with ether. The reaction step is shown in Scheme 2.4.

The overall yield of the reaction was 65%. The melting point of ADN recrystallised from dioxane-ethyl acetate mixture (5:1) is in the range of 89°C–94°C.

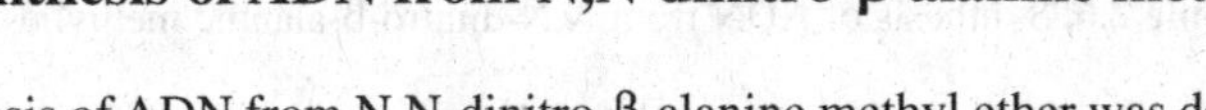

$$(O_2N)_2NCH_2CH_2CN \xrightarrow{NH_3} NH_4{}^+[N(NO_2)_2]^- + CH_2=CH\text{-}CN$$

Scheme 2.4. Synthesis of ADN from N,N-dinitro-β-aminopropionitrile.

2.5. Synthesis of ADN from N,N-dinitro-β-alanine methyl ester

The synthesis of ADN from N,N-dinitro-β-alanine methyl ether was described by Lukyanov *et al.*[6] The reaction involves many steps. To a solution of methyl chloroformate in ether, a suspension of β-alanine methyl ester hydrochloride in CHCl$_3$ was added in portions. A solution of triethylamine (Et$_3$N) was then added at 20°C–25°C and stirred. The formed Et$_3$N.HCl was filtered off and the distillation of filtrate gave N-methoxycarbonyl-β-alanine methyl ester. The methyl ester was then treated with 99% HNO$_3$ and stirred for some time and poured to crushed ice. The resulting precipitate was filtered, washed with water and dried over P$_2$O$_5$ in a vacuum desiccator to give N-methoxycarbonyl-N-nitro-β-alanine methyl ester. The ammonium salt of N-nitro-β-alanine methyl ester was prepared by passing NH$_3$ at 0°C to a solution of N-nitro-β-alanine methyl ester in dry ether.

The resulting precipitate was filtered and washed with ether and dried. The ammonium salt of N-nitro-β-alanine methyl ester was treated with NO_2BF_4 at −10°C under vigorous stirring in dry CH_3CN. The reaction mixture was then poured into water and extracted with CH_2Cl_2 and dried over $MgSO_4$. Evaporation of the solvent gave N,N-dinitro-β-alanine methyl ester. The reaction steps are shown in Scheme 2.5.

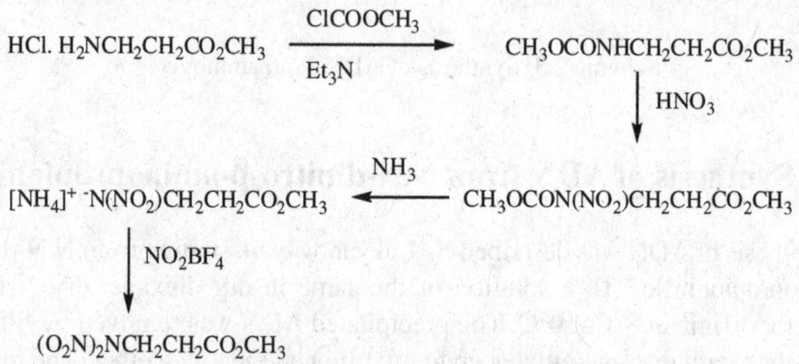

$$HCl.\ H_2NCH_2CH_2CO_2CH_3 \xrightarrow[Et_3N]{ClCOOCH_3} CH_3OCONHCH_2CH_2CO_2CH_3$$

$$\downarrow HNO_3$$

$$[NH_4]^+\ ^-N(NO_2)CH_2CH_2CO_2CH_3 \xleftarrow{NH_3} CH_3OCON(NO_2)CH_2CH_2CO_2CH_3$$

$$\downarrow NO_2BF_4$$

$$(O_2N)_2NCH_2CH_2CO_2CH_3$$

Scheme 2.5. Synthesis of N,N-dinitro-β-alanine methyl ester.

To a solution of N,N-dinitro-β-alanine methyl ester in dry dioxane, NH_3 was passed at 5°C–8°C. The precipitated ADN was filtered and washed with dioxane. The reaction is shown in Scheme 2.6. The yield of the reaction was 48.9%.

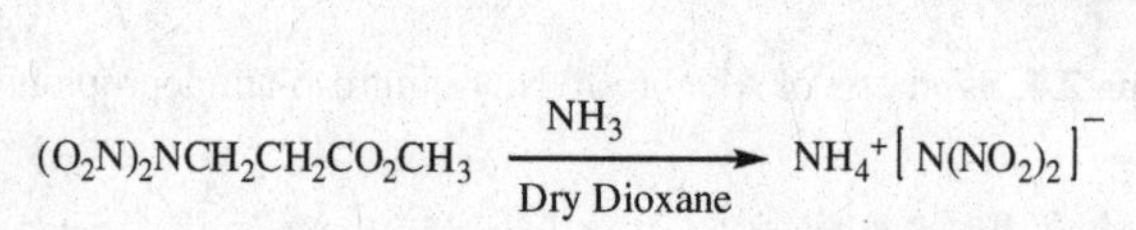

$$(O_2N)_2NCH_2CH_2CO_2CH_3 \xrightarrow[Dry\ Dioxane]{NH_3} NH_4^+ \left[N(NO_2)_2 \right]^-$$

Scheme 2.6. Synthesis of ADN from N,N-dinitro-β-alanine methyl ester.

2.6. Synthesis of ADN from N,N-dinitro Derivatives of Organic Amides

ADN can be synthesised from ammonium salt of N-nitrobenzamide using NO_2BF_4.[7] The nitrating agent and the ammonium salt of N-nitrobenzamide in dry CH_3CN were allowed to stir at −65°C. The residue after the reaction was separated and treated with NH_3 to give ADN. The yield of the reaction was 61%. The reaction steps are shown in Scheme 2.7.

ADN was also obtained by the nitration of ammonium salt of N-nitro-p-toluenesulfamide with nitronium tetrafluoborate. The reaction was carried out in CH_3CN and abs.EtOAc at −65°C. The mixture after the reaction was filtered.

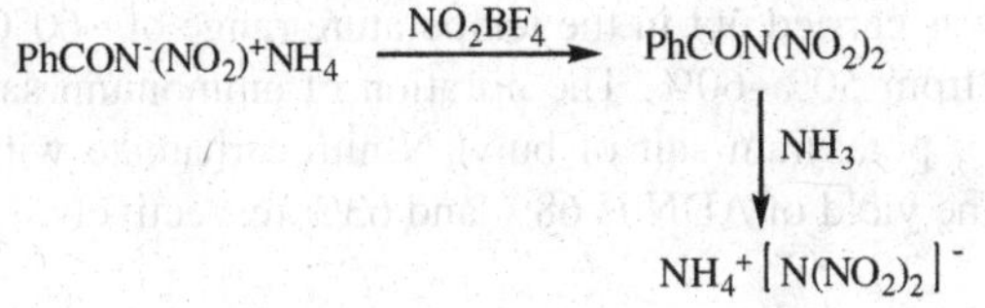

$$PhCON^-(NO_2)^+NH_4 \xrightarrow{NO_2BF_4} PhCON(NO_2)_2$$

$$\downarrow NH_3$$

$$NH_4^+ \left[N(NO_2)_2 \right]^-$$

Scheme 2.7. Synthesis of ADN from ammonium salt of N-nitrobenzamide.

Liquid NH_3 was added to the filtrate and the formed residue was filtered off and the filtrate was evaporated to give ADN. The reaction is shown in Scheme 2.8. The overall yield of the reaction was 55%.

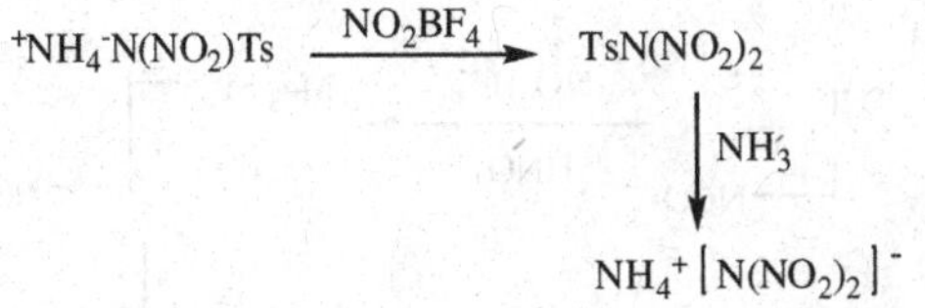

$$^+NH_4^-N(NO_2)Ts \xrightarrow{NO_2BF_4} TsN(NO_2)_2$$

$$\downarrow NH_3$$

$$NH_4^+ \left[N(NO_2)_2 \right]^-$$

Scheme 2.8. Synthesis of ADN from ammonium salt of N-nitro-p-toluenesulfamide.

2.7. Synthesis of ADN from Alkyl Carbamate

The reaction of an alkyl carbamate with acetic anhydride and HNO_3 at 0°C forms N-(alkoxycarbonyl)-N-nitroamide (nitrourethane). Treatment of the latter with ammonium acetate forms ammonium nitrourethane which upon further nitration with NO_2BF_4 gave dinitrourethane. The base treatment of the latter with ammonia gives ADN.[8] as illustrated in Scheme 2.9.

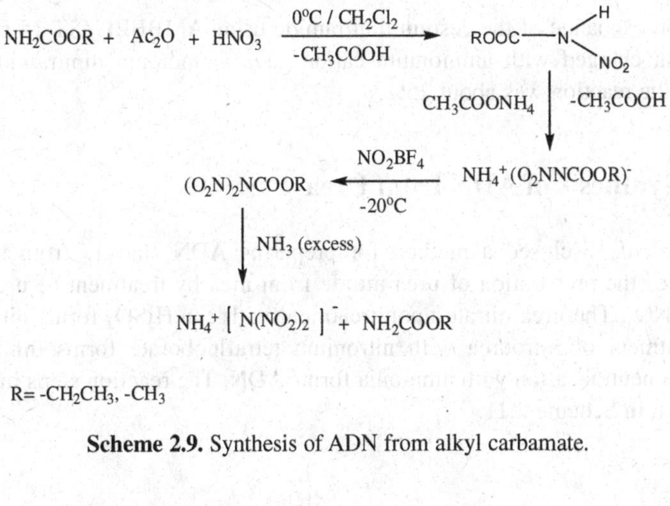

$$NH_2COOR + Ac_2O + HNO_3 \xrightarrow[-CH_3COOH]{0°C / CH_2Cl_2} ROOC-N\overset{H}{\underset{NO_2}{\big\backslash}}$$

$$CH_3COONH_4 \downarrow -CH_3COOH$$

$$(O_2N)_2NCOOR \xleftarrow[-20°C]{NO_2BF_4} NH_4^+(O_2NNCOOR)^-$$

$$\downarrow NH_3 \text{ (excess)}$$

$$NH_4^+ \left[N(NO_2)_2 \right]^- + NH_2COOR$$

R= -CH$_2$CH$_3$, -CH$_3$

Scheme 2.9. Synthesis of ADN from alkyl carbamate.

The reaction was carried out in the temperature range of $-60°C$ to $-20°C$ and the yield varies from 50%–60%. The nitration of ammonium salt of propyl N-nitrocarbamate or potassium salt of butyl N-nitrocarbamate with NO_2BF_4 also affords ADN.[7] The yield of ADN is 68% and 63% respectively.

2.8.   Synthesis of ADN from 2-(trimethylsilyl-ethyl)-1-isocyanate

Reaction of 2-(trimethylsilyl-ethyl)-1-isocyanate with NO_2BF_4 in the presence of HNO_3 yields 1-(N,N-dinitramino)-2-(trimethylsilyl)ethane which on treatment with CsF formed cesium dinitramide.[1a] The reaction steps are shown in Scheme 2.10.

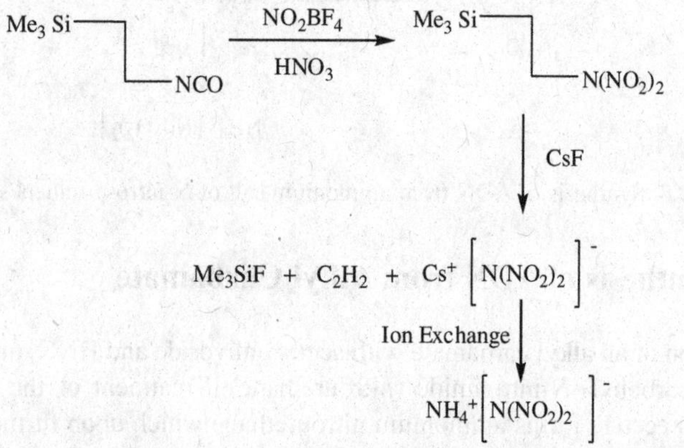

Scheme 2.10. Synthesis of ADN from 2-(trimethylsilyl-ethyl)-1-isocyanate.

The ion exchange of the cesium dinitramide using AMBERLYST 15 sulfonic acid resin charged with ammonium cation gave ammonium dinitramide. The yield of the reaction was about 25%.

2.9.   Synthesis of ADN from Urea

Hatano *et al.* disclosed a method for preparing ADN starting from urea.[9,10] It involves the preparation of urea nitrate from urea by treatment of urea with dilute HNO_3. The urea nitrate upon treatment with con.H_2SO_4 forms nitrourea. The treatment of nitrourea with nitronium tetrafluoborate forms dinitrourea which on neutralisation with ammonia forms ADN. The reaction steps involved are shown in Scheme 2.11.

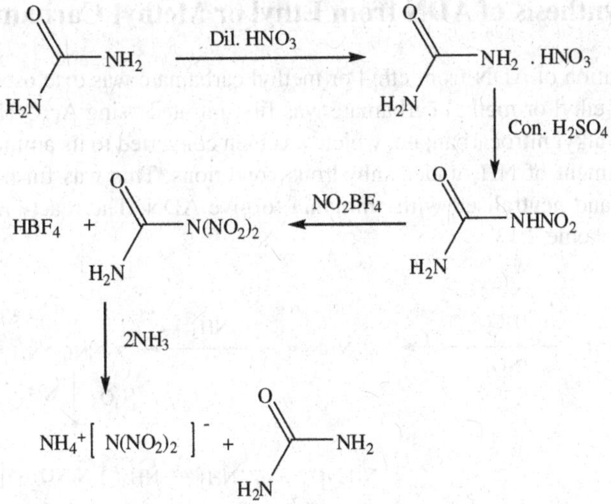

Scheme 2.11. Synthesis of ADN from urea.

2.10. Synthesis of ADN from bis(2-cyanoethyl) amine

The preparation of ADN from bis(2-cyanoethyl) amine was disclosed by Kyoo-Hyun Chung *et al.*[11] Nitration of bis(2-cyanoethyl) amine using HNO_3/Ac_2O mixture followed by decyanoethylation by a base and further nitration with NO_2BF_4 gave ADN. The reaction steps are shown in Scheme 2.12. The reaction was carried out at $-10°C$ and the yield of ADN was 60%.

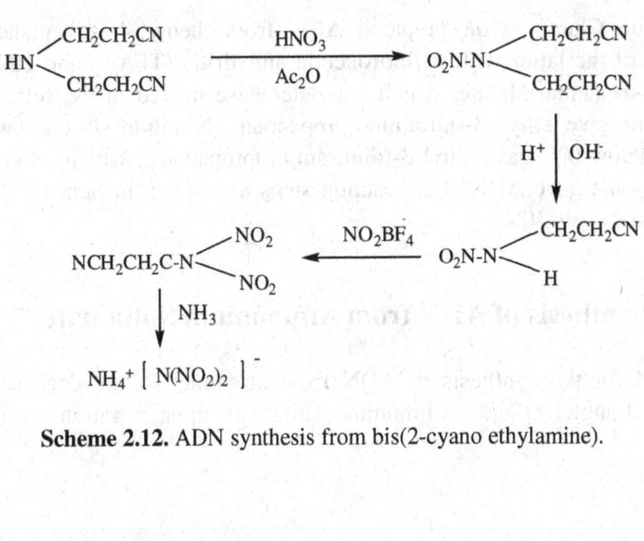

Scheme 2.12. ADN synthesis from bis(2-cyano ethylamine).

2.11.   Synthesis of ADN from Ethyl or Methyl Carbamate

The preparation of ADN from ethyl or methyl carbamate was disclosed by Stern *et al.*[12] The ethyl or methyl carbamate was first nitrated using Ac_2O/HNO_3 mixture to give alkyl nitrocarbamate, which was then converted to its ammonium salt by the treatment of NH_3 under anhydrous conditions. This was further nitrated with N_2O_5 and neutralised with ammonia to give ADN. The reaction steps are shown in Scheme 2.13.

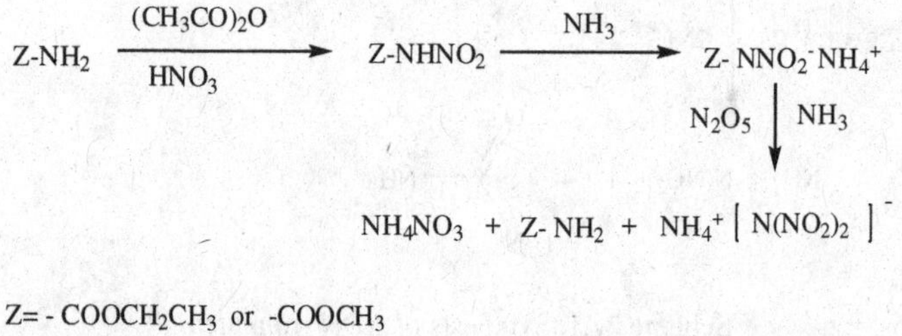

$$Z\text{-}NH_2 \xrightarrow[HNO_3]{(CH_3CO)_2O} Z\text{-}NHNO_2 \xrightarrow{NH_3} Z\text{-}NNO_2^- NH_4^+$$

$$\downarrow {\scriptstyle N_2O_5 \mid NH_3}$$

$$NH_4NO_3 \ + \ Z\text{-}NH_2 \ + \ NH_4^+ \left[N(NO_2)_2 \right]^-$$

$$Z = \text{-}COOCH_2CH_3 \text{ or } \text{-}COOCH_3$$

Scheme 2.13. Synthesis of ADN from ethyl or methyl carbamate.

The first step of nitration was performed at 0°C, while the second step of the reaction was performed in CH_2Cl_2 medium at −48°C. The overall yield of the reaction was about 70%–76%.

2.12.   Synthesis of ADN from Diethyl-4-azaheptanedioate

Kyoo-Hyun Chung *et al.* prepared ADN from diethyl-4-azaheptanedioate.[13] Nitration of the latter with trifluoroacetic anhydride (TFAA) and HNO_3 gave N-nitro-4-azaheptanedioate, which was later base treated in the retro-Michael reaction to give ethyl 3-(nitramino)propanoate. Nitration of the latter with NO_2BF_4 below 0°C gave ethyl 3-(dinitramino)propanoate, which upon reaction with ammonia gave ADN. The reaction steps are shown in Scheme 2.14. The yield of ADN was 30%.

2.13.   Synthesis of ADN from Ammonium Sulfamate

A method for the synthesis of ADN from ammonia-sulfate derivatives was given by Langlet *et al.*[14] Ammonium sulfamate upon nitration with fuming

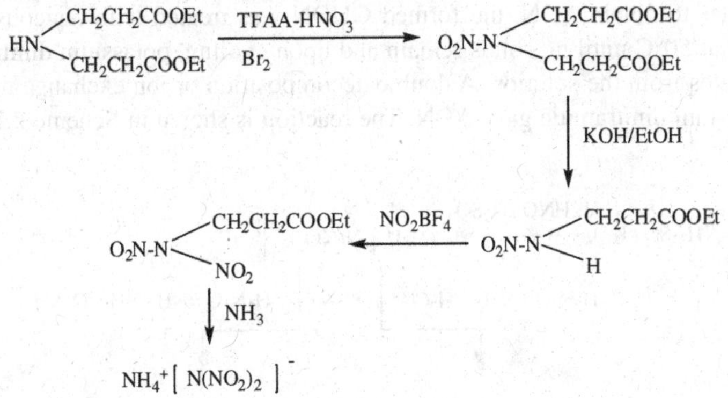

Scheme 2.14. Synthesis of ADN from diethyl-4-azaheptanedioate.

HNO_3/con.H_2SO_4 mixture at $-40°C$ followed by dilution with water and neutralisation with NH_3 gave ADN. The reaction step is shown in Scheme 2.15.

The overall yield of ADN was about 60%. The use of NH_2NO_2 or $NH_4NH_2CO_2$ or NH_2SO_3H or $NH(SO_3H)_2$ or $N(SO_3H)_3$ and its salts with various cations as starting materials in the above reaction also yields ADN. An adsorption procedure using activated charcoal was suggested for the removal of ADN from the neutralised solution. Santhosh *et al.* have performed a detailed analysis on the adsorption of ADN from aqueous solutions using powdered activated charcoal.[15] The elution was achieved using hot water as eluent and the equilibrium concentrations and adsorption isotherms were derived.

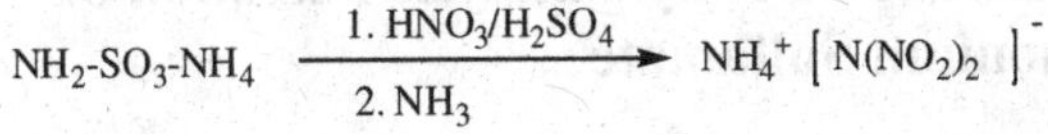

$$NH_2\text{-}SO_3\text{-}NH_4 \quad \xrightarrow[\text{2. } NH_3]{\text{1. } HNO_3/H_2SO_4} \quad NH_4^+ \left[N(NO_2)_2 \right]^-$$

Scheme 2.15. Synthesis of ADN from ammonium sulfamate.

2.14. Synthesis of ADN from Potassium Sulfamate using Cyanoguanidine or Guanylurea as Neutralising Agents

Carin *et al.* disclosed a process for the preparation of ADN starting from salts of sulfamic acid.[16] In a typical reaction, potassium sulfamate was nitrated using fuming HNO_3/con.H_2SO_4 mixture. The mixture after the reaction was treated with either cyanoguanidine or guanylurea to give guanylurea dinitramide (GUDN). The precipitated GUDN was removed by filtration and dried in vacuum.

In order to obtain ADN, the formed GUDN was treated with aqueous KOH solution at 50°C until no solids remain and upon cooling, potassium dinitramide precipitates from the solution. A double decomposition or ion exchange reaction of potassium dinitramide gave ADN. The reaction is shown in Scheme 2.16.

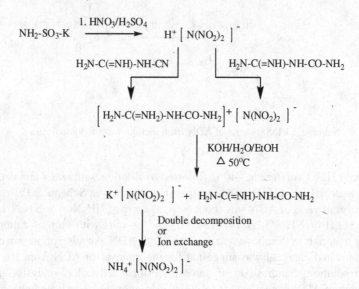

Scheme 2.16. Synthesis of ADN from potassium sulfamate using cyanoguanidine or guanylurea as neutralising agents.

2.15. Synthesis of ADN by Catalytic Nitration of Ammonium Sulfamate

Choudary *et al.* describe a process for the preparation of ADN by catalytic nitration in the presence of fuming HNO_3 and solid acid catalysts.[17] The catalysts are selected from a group consisting of montmorillonite clay, metal ion exchanged K10 montmorillonite clay and surface supported catalysts.

Fe^{3+}-exchanged montmorillonite catalyst, titanium supported on silica-alumina (TISIAL), molybdenum supported on silica-alumina (MOSIAL) and acid treated TISIAL and MOSIAL were used as catalysts. The reaction was carried out by mixing fuming HNO_3 and the catalyst and cooling the mixture to −40°C, ammonium sulfamate was then added in portions. The reaction mixture was diluted with water and neutralised with ammonia to give ADN. The reaction is shown in Scheme 2.17. The yield of the reaction was 7%–23%.

$$NH_2\text{-}SO_3\text{-}NH_4 \xrightarrow[\text{2. NH}_3]{\text{1. HNO}_3 \text{ / Solid Acid Catalysts}} NH_4^+ \left[\; N(NO_2)_2 \;\right]^-$$

Scheme 2.17. Catalytic nitration of ammonium sulfamate.

A summary of the starting materials, nitrating agents and experimental conditions for the synthesis of ADN is given in Table 2.1.

Table 2.1. Starting Materials and Nitrating Agents for the Synthesis of ADN.

Starting material	Nitrating agent	Temperature (°C)	Yield (%)	Reference
$Me_3Si(CH_2)_2NCO$	NO_2BF_4/HNO_3	<0	25	[1a]
NH_2NO_2	a.NO_2BF_4	−20	60	[1a]
	b.N_2O_5	−20	<1	[1a]
	c.$NO_2HS_2O_7$	−40	53	[1a]
	d.NO_2SO_3F	−40	85 (5min)	[1b]
	e.$(NO_2)_2S_2O_7$	−40	49 (90min)	[1b]
NH_4COONH_2	NO_2BF_4	0	15	[3]
NH_3	a.NO_2BF_4	−78	25	[4]
	b.N_2O_5	−78	15	[4]
	c.$NO_2HS_2O_7$	−78	15	[4]
$NH_2CONHNO_2$	NO_2BF_4	−40	20	[9,10]
$NH_2COOC_2H_5$	a.N_2O_5	−48	70	[8]
	b.NO_2BF_4	−60	60	[8]
$HN(CH_2CH_2COOEt)_2$	NO_2BF_4	<0	31	[13]
$HN(CH_2CH_2CN)_2$	NO_2BF_4	−10	60	[11]
$NH_2SO_3NH_4$	HNO_3/H_2SO_4	−45	60	[14]
[$NH_2SO_3K/$ cyanoguanidine/ guanylurea]	HNO_3/H_2SO_4	−40−25	—	[16]
$NH_2SO_3NH_4$	HNO_3/solid acids	−40	7–23	[17]
$NH_4N(NO_2)$ $CH_2CH_2CO_2CH_3$	NO_2BF_4	−10	48.9	[6]
$Ph\,CON(NO_2)NH_4$	NO_2BF_4	−65	61	[7]
$NH_4N(NO_2)Ts$	NO_2BF_4	−65	55	[7]
$NH_4N(NO_2)CO_2Pr$	NO_2BF_4	−40	68	[7]
$KN(NO_2)CO_2Bu$	NO_2BF_4	−40	63	[7]

References

1. (a) JC Bottaro, RJ Schmitt, PE Penwell, *et al.*, Dinitramide salts and method of making same, *U.S. Patent 5254324*, 1993, SRI International, USA. (b) OA Lukyanov, SN Shvedova, EV Shepelev, ON Varfolomeeva, NN Malkina, VA Tartakovsky, Dinitramide and its salts 11. Synthesis of dinitramide by nitration of nitramide with nitryl salts, *Russ Chem Bull* **45**(6): 1497–1498, 1996.
2. M Malesa, W Skupinski, M Jamroz, Separation of ammonium dinitramide from reaction mixture, *Prop Expl Pyro* **24**: 83–89, 1999.
3. JC Bottaro, RJ Schmitt, PE Penwell, *et al.*, Method of forming dinitramide salts, *U.S. Patent 5198204*, SRI International, USA, 1993.
4. RJ Schmitt, JC Bottaro, PE Penwell, *et al.*, Process for forming ammonium dinitramide salt by reaction between ammonia and a nitronium containing compound, *U.S. Patent 5316749*, SRI International, USA, 1994.
5. OA Lukyanov, VP Gorelik, VA Tartakovskii, Dinitramide and its salts 1. Synthesis of dinitramide salts by decyanoethylation of N,N-dinitro-β-aminopropionitrile, *Russ Chem Bull* **43**(1): 89–92, 1994.
6. OA Lukyanov, Yu V Konnova, TA Klimova, *et al.*, Dinitramide and its salts 2. Dinitramide in Michael and retro-Michael-type reactions, *Russ Chem Bull* **43**(7): 1200–1202, 1994.
7. OA Lukyanov, IK Kozlova, OP Shitov, *et al.*, Dinitramide and its salts 10. Synthesis of dinitramide salts from N,N-dinitro derivatives of organic amides, *Russ Chem Bull* **45**(4): 863–867, 1996.
8. RJ Schmitt, JC Bottaro, PE Penwell, *et al.*, Process for forming a dinitramide salt or acid by reaction of a salt or free acid of an N(alkoxycarbonyl)N-nitroamide with a nitronium-containing compound followed by reaction of the intermediate product respectively with a base or alcohol, *U.S. Patent 5415852*, SRI International, USA, 1995.
9. H Hatano, T Onda, K Shiino, *et al.*, New synthetic method and properties of ammonium dinitramide, *Kayaku Gakkaishi*, **57**(4): 160–165, 1996.
10. S Suzuki, S Miyazaki, H Hatano, *et al.*, Synthetic method for forming ammonium dinitramide (ADN), *U.S. Patent 5659080*, Nissan Motor Company and Hosaya Fireworks Company, Japan, 1997.
11. Kyoo-Hyun Chung, Hyun-Ho Sim, Study on the synthesis of ammonium dinitramide, *J Korean Chem Soc* **41**(12): 661–665, 1997.
12. AG Stern, WM Koppes, ME Sitzmann, *et al.*, Process for preparing ammonium dinitramide, *U.S. Patent 5714714*, Secretary of the Navy, USA, 1998.
13. Kyoo-Hyun Chung, Hyun-Ho Sim, A study on the synthesis of dinitramide salts, *J Korean Ind & Engg Chem* **9**(1): 155–157, 1998.
14. A Langlet, H Ostmark, N Wingborg, Method of preparing dinitramidic acid and salts thereof, *U.S. Patent 5976483*, Forsvarets Forskningsanstalt, Sweden, 1999.

15. G Santhosh, S Venkatachalam, KN Ninan, *et al.*, Adsorption of ammonium dinitramide (ADN) from aqueous solutions 1. Adsorption on powdered activated charcoal, *J Haz Mater* **B98:** 117–126, 2003.

16. Vorde Carin, Skifs Henrik, Method of producing salts of dinitramidic acid, *WO2005/070823A1*, 2005.

17. (a) BM Choudary, M Lakshmi Kantam, K Jeeva Ratnam, *et al.*, Process for the preparation of dinitramidic acid and salts thereof, *U.S. Patent 6787119*, 2004 & *European Patent 1344748 A1*, 2003. (b) BM Choudary, M Lakshmi Kantam, K Jeeva Ratnam, *et al.*, A process for the preparation of dinitramidic acid and salts, *Ind. Patent Appl. No. 157/DEL/2002*, 2006.

Chapter 3

SYNTHESIS OF OTHER DINITRAMIDE SALTS

A plethora of dinitramide salts apart from ADN have been synthesised and characterised. These salts may find use in organic synthesis, pyrotechnic composition, gas generators or civil and military applications. This chapter reviews the different kinds of organic, inorganic, metal and complex salts of dinitramide.

3.1. General Methods of Preparation

The preparation of dinitramide salts fall under one of the following methods.

[A] The interaction of metal hydroxides with β-substituted alkyl-N, N-dinitramines in organic solvents gives metallic dinitramide salts as shown in Equation 3.1.

$$X\text{-}CH_2\text{-}CH_2N(NO_2)_2 \ + \ MOH \ \xrightarrow[\text{-XCH=CH}_2]{} \ M\text{-}N(NO_2)_2 + H_2O \qquad (3.1)$$

Where, X = -CN, -CHO, -COR, -COOR and M = K, Rb, Cs

[B] Cation exchange reaction of Ag, K or Cs salt of dinitramide with an appropriate metal halide in water or organic solvents gives the corresponding metal dinitramide salt as shown in Equation 3.2.

$$n \text{ M-N(NO}_2)_2 + \text{M'X}_n \longrightarrow \text{M'(N(NO}_2)_2)_n + n\text{MX} \qquad (3.2)$$

Where, $M = Ag, K, Cs$ and $X = Br, Cl, F$

[C] The treatment of ammonium salt of dinitramide with solutions of strong bases or the reaction of aqueous and nonaqueous dinitramidic acid with an appropriate metal hydroxide, oxide or carbonate will give metal dinitramide salts as shown in Equation 3.3.

$$n \text{ NH}_4\text{N(NO}_2)_2 + \text{M(OH)}_n \longrightarrow \text{M(N(NO}_2)_2)_n + n\text{ NH}_3 + n\text{ H}_2\text{O}$$

$$2n \text{ HN(NO}_2)_2 \xrightarrow{\begin{array}{c} \text{M}_2\text{O}_n \text{ or } 2\text{M(OH)}_n \\ \text{or } \text{M}_2(\text{CO}_3)_n \end{array}} 2\text{M(N(NO}_2)_2)_n \qquad (3.3)$$

[D] The reaction of ADN or dinitramidic acid with excess of organic amine either in bulk or in a solvent at 60°C for 4 hours–5 hours will give the desired dinitramide salt as shown in Equation 3.4.

$$\text{R-NH}_2 + \text{HN(NO}_2)_2 \longrightarrow \text{R-N(NO}_2)_2 + \text{NH}_3$$

$$\text{R-NH}_2 + \text{NH}_4\text{N(NO}_2)_2 \longrightarrow \text{R-N(NO}_2)_2 + 2\text{NH}_3 \qquad (3.4)$$

[E] ADN or a metal dinitramide was heated with equimolar quantity of carbonate, hydrochloride or nitrate salt of an amine in isopropanol or ethanol till the evolution of NH_3 or CO_2 is ceased. Evaporation of the solvent affords the corresponding dinitramide salt. The reactions are shown in Equation 3.5.

$$\text{R-NH}_2.\text{H}_2\text{CO}_3 + \text{NH}_4\text{N(NO}_2)_2 \longrightarrow \text{R-NH}_2.\text{HN(NO}_2)_2 + \text{NH}_3 + \text{CO}_2 + \text{H}_2\text{O}$$

$$\text{R-NH}_2.\text{HCl} + \text{MN(NO}_2)_2 \longrightarrow \text{R-NH}_2.\text{HN(NO}_2)_2 + \text{MCl}$$

$$\text{R-NH}_2.\text{HNO}_3 + \text{MN(NO}_2)_2 \longrightarrow \text{R-NH}_2.\text{HN(NO}_2)_2 + \text{MNO}_3 \qquad (3.5)$$

Where, $M = Ag, K$

[F] The reaction of ADN with equimolar quantity of Ba(OH)_2 in aqueous medium gives barium dinitramide. The reaction of the same with a sulphate salt of an amine in ethanol gives the corresponding amine dinitramide salt as shown in Equation 3.6.

$$\text{Ba(OH)}_2 + 2\text{NH}_4\text{N(NO}_2)_2 \longrightarrow \text{Ba(N(NO}_2)_2)_2 + 2\text{NH}_3 + 2\text{H}_2\text{O}$$

$$2\text{RNH}_2.\text{H}_2\text{SO}_4 + \text{Ba(N(NO}_2)_2)_2 \longrightarrow 2\text{RNH}_2.\text{HN(NO}_2)_2 + \text{BaSO}_4 \qquad (3.6)$$

3.2. Metal Dinitramide Salts

Mono and divalent metals of group I, II, VII and VIII form stable salts of dinitramide. The metal dinitramide salts are crystalline with low melting or decomposition points. The salts of lithium and other heavy metals are soluble in low polarity solvents such as diethyl ether. The potassium and cesium salts of dinitramidic acid are insensitive to mechanical action, but salts of heavy metals are more sensitive to impact and friction.

The silver salt of dinitramide is capable of forming complexes with solvents such as MeCN or dioxane. $Fe(N_3O_4)_2.7H_2O$ oxidizes and decomposes readily in the presence of air. The Na, K, Rb and Cs dinitramide salts may find applications in military and civilian pyrotechnics. The metal dinitramide salts, apart from providing energy for the composition, provide pyrotechnic effects and increase the density of the formulation. The methods of preparation of various metal dinitramide salts, their melting and decomposition temperatures are summarised in Table 3.1.

Table 3.1. Metal Dinitramide Salts

Dinitramide	Method of preparation	Melting point (°C)	Reference
Li $N(NO_2)_2.H_2O$	[B]	68–73	[1]
Li $N(NO_2)_2$	[B]	158 (decomp)	[1]
Na $N(NO_2)_2$	[B]	101–107	[1]
K $N(NO_2)_2$	[A, C]	127–131	[2]
Rb $N(NO_2)_2$	[A, C]	102–106	[2]
Cs $N(NO_2)_2$	[A, C]	85–88	[2]
Ag $N(NO_2)_2$	[A, C]	125–131 (decomp)	[1]
Pb $N(NO_2)_2$	[A, C]	102–106	[3]
Ba $(N_3O_4)_2.H_2O$	[F]	74–76	[1]
Ni $(N_3O_4)_2.6H_2O$	[B]	80–83 (decomp)	[1]
Ni $(N_3O_4)_2.2H_2O$	[B]	93–97 (decomp)	[1]
Mg $(N_3O_4)_2.6H_2O$	[B]	89–93 (decomp)	[1]
Mg $(N_3O_4)_2.3H_2O$	[B]	60–65 (decomp)	[1]
Cu $(N_3O_4)_2.3H_2O$	[B]	51–56	[1]
Cu $(N_3O_4)_2. H_2O$	[B]	—	[1]
Fe $(N_3O_4)_2.7H_2O$	[B]	85 (decomp)	[1]
Co $(N_3O_4)_2.6H_2O$	[B]	82–86	[1]
Mn $(N_3O_4)_2.8H_2O$	[B]	41–63	[1]
Hg $(N_3O_4)_2$	[C]	93–103 (decomp)	[4]

3.3. Organic Dinitramide Salts

Numerous dinitramide salts bearing an organic moiety were synthesised and characterised. Most of the salts were prepared by procedures outlined in Section 3.1 with a few exceptions. Their melting points along with the method of preparation are summarised in Table 3.2.

Table 3.2. Organic Dinitramide Salts

Dinitramide	Method of preparation	Melting point (°C)	Reference
1,5-diamino-4-methyltetrazolium dinitramide[a]	—	85	[5]
1,2,3-triazolium dinitramide	[D]	61	[6a]
1,2,4-triazolium dinitramide	[D]	75	[6a]
2-hydroxyethylhydrazinium monodinitramide	[C]	—	[6b]
3,3-dinitroazetidinium dinitramide[b]	—	139	[7]
3,5-dimethylpyridine dinitramide	[F]	70–73	[8]
3,6-dihydrazino-1,2,4,5-tetrazine bis(dinitramide)[c]	—	—	[9]
4-amino-1,2,4-triazolium dinitramide	[D]	20	[6a]
5-aminotetrazole dinitramide	[F]	83–86	[8]
Aminoguanidine dinitramide	[E]	92–94	[10]
Aniline dinitramide	[D]	99–100	[10]
Benzylamine dinitramide	[D]	59–61	[8]
Biguanidinium mono-dinitramide	[F]	—	[11]
Biguanidinium bis-dinitramide	[F]	126–129	[11]
Bis(cyanoethyl)amine dinitramide	[D]	115–117	[10]
Butane diammonium dinitramide	[D]	67–73	[12]
Cubane 1,4-bis (ammonium dinitramide)[d]	—	No melting	[13]
Cubane 1,2,4,7-tetrakis(ammonium dinitramide)[d]	—	No melting	[13]
Cyanoethylamine dinitramide	[D]	67–69	[10]
Diethanolamine dinitramide	[D]	75–78	[8]
Dimethylamine dinitramide	[D]	31–33	[10]
Dimethylhydrazine dinitramide	[E]	112–118	[10]
Ethane-1,2-diammonium dinitramide	[E]	123–126	[10,14]
Ethanolamine dinitramide	[D]	37–39	[8]
Ethylene bis(oxyamine) monodinitramide[e]	—	57–59	[15]
Ethylene bis(oxyamine)bis (dinitramide)[e]	—	<0	[15]
Guanidine dinitramide	[E]	135–139	[10]
Hexamethylenediamine bisdinitramide	[D]	89–90	[8]

(Continued)

Table 3.2. (*Continued*)

Dinitramide	Method of preparation	Melting point (°C)	Reference
Melaminium dinitramide	[D]	—	[16]
Methylamine dinitramide	[D]	43–47	[10]
Methylene bis(oxyamine) monodinitramide[f]	—	96	[17]
Methylene bis(oxyamine)bis (dinitramide)[f]	—	85–95	[17]
Methylhydroxylamine dinitramide	[E]	oil	[10]
m-nitroaniline dinitramide	[D]	101–103	[8]
Morpholine dinitramide	[D]	82–84	[8]
N-guanylurea dinitramide	[F]	No melting	[18]
Triethylenetetramine tris(dinitramide)	[D]	Oil	[19]
Triethylenetetramine tetrakis(dinitramide)	[D]	159	[19]
Trimethylamine dinitramide	[D]	100–128 (decomp)	[10]
N-methyl urotropinium dinitramide[g]	—	121–124	[20]
N-nitropyridinium dinitramide[h]	—	55–58	[10]
o-Toluidine dinitramide	[F]	70–71	[8]
Piperazine bisdinitramide	[D]	212–214	[8]
Tetramethylammonium dinitramide[i]	—	228	[10]
Triaminoguanidine dinitramide (TAGDN)[j]	—	85	[21]
Triazidocarbenium dinitramide[k]	—	—	[22]
Urea dinitramide	[E]	98–100	[10]

[a] by reaction of 1,5-diamino-4-methyl-1H-tetrazolium iodide with silver dinitramide in MeCN,
[b] neutralisation of dinitramidic acid with 3,3-dinitroazetidine in methanol,
[c] neutralisation of dinitramidic acid with 3,6-dihydrazino-1,2,4,5-tetrazine in water,
[d] ion exchange reaction of the corresponding perchlorate salts,
[e] reaction of ethylene bis(oxyamine) in methanol with dinitramidic acid,
[f] reaction of methylene bis(oxyamine) in methanol with dinitramidic acid,
[g] reaction of N-methylurotropinium iodide with $[AgNCCH_3][N_3O_4]$,
[h] reaction of $(C_5H_5NNO_2)^+ BF_4^-$ and potassium dinitramide,
[i] reaction of tetramethylammonium bromide with silver dinitramide,
[j] reaction of $C(NH_2)_3N_3O_4$ with hydrazine-hydrate in boiling dioxane,
[k] reaction of $C(N_3)_3^+ BF_4^-$ and potassium dinitramide.

The amine salts of dinitramide posses good thermal stability and low susceptibility to hydrolysis and can find use in explosives or components of energetic formulations. The triethylenetetramine dinitramide salts are not susceptible to
hydrolysis. Quarternary ammonium and hydrazinium salts have high melting and
decomposition points. The quarternary, aminoguanidine, acetamidine dinitramide
salts were nonhygroscopic, while the methylamine, dimethylamine salts were
hygroscopic. The methoxyamine dinitramide is readily soluble in nonpolar

solvents. Azetidine-based dinitramide salts are expected to have high performance because of the associated high strain energy with the four membered ring. The 3,3-dinitroazetidinium cation has a high oxygen balance and is a useful substitute for the conventional cations such as ammonium, guanidine, hydrazine etc. The triazidocarbenium dinitramide was marginally stable at room temperature and is extremely sensitive to impact and friction. The cubane dinitramide salts also have high strain energy and thus may have potential application in advanced propellants. The guanylurea dinitramide is stable up to 200°C, is not hygroscopic, is insensitive and not soluble in cold water. This salt is currently being investigated for use in automotive air bags. Its exhaust is free of toxic, corrosive and flammable gases are attractive in terms of passenger safety.

3.4.　Inorganic Dinitramide Salts

Inorganic dinitramide salts viz. hydrazinium dinitramide and hydroxylamine dinitramide have been reported. The method of preparation and the melting points for these salts are given in Table 3.3.

Due to the high sensitive nature of hydrazinium dinitramide, its use in missiles and propellants has its own limitations. The hydroxylamine dinitramide is readily soluble in nonpolar solvents such as ether and its low melting point is rather disadvantageous for many applications. However, these inorganic dinitramide salts may find use as monopropellants.

3.5.　Metal Complexes of Dinitramide

Dinitramide salts form molecular complexes with many ligands. Ammine complexes are formed when ammonia is passed through a solution containing metal dinitramide. Complexes with nitrogen containing ligands such as pyridine and morpholine were also prepared. Trammel *et al.* reported a rhenium metal complex with bipyridyl and carbonyl units and identified its structure by X-ray methods. The reaction of a reactive precursor fac-Re(bpy)-(CO)$_3$OSO$_2$CF$_3$ with potassium dinitramide in water at room temperature gave fac-Re(bpy)-(CO)$_3$N$_3$O$_4$ in 75%

Table 3.3. Inorganic Dinitramide Salts

Dinitramide	Method of preparation	Melting point °C	Reference
NH$_2$NH$_2$.HN$_3$O$_4$	[C]	77–80	[10]
H$_2$NOH.HN$_3$O$_4$	[E]	18–23	[10]

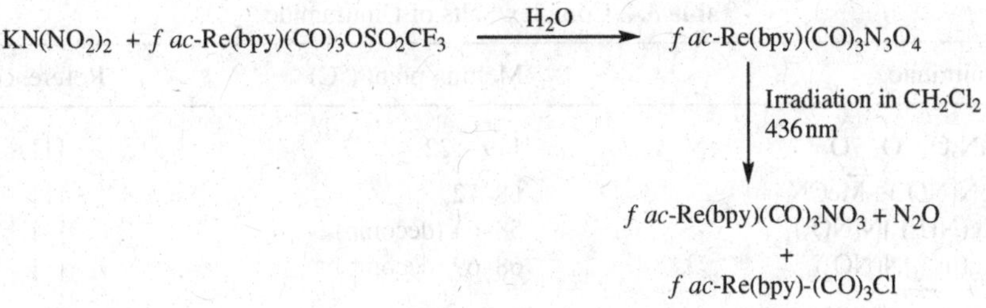

$$KN(NO_2)_2 + f \, ac\text{-}Re(bpy)(CO)_3OSO_2CF_3 \xrightarrow{H_2O} f \, ac\text{-}Re(bpy)(CO)_3N_3O_4$$

Irradiation in CH_2Cl_2
436 nm

$$\downarrow$$

$$f \, ac\text{-}Re(bpy)(CO)_3NO_3 + N_2O$$
$$+$$
$$f \, ac\text{-}Re(bpy)\text{-}(CO)_3Cl$$

Scheme 3.1. Preparation and decomposition of Re-complex of dinitramide.

yield.[23] The X-ray crystal structure indicates that the coordination occurs at the central nitrogen atom of the dinitramide.

The preparation and the decomposition steps of the complex are shown in Scheme 3.1. The authors have studied the photosensitivity of the complex in solution and identified *fac*-Re(bpy)-(CO)$_3$NO$_3$ as the product formed along with trace amounts *fac*-Re(bpy)-(CO)$_3$Cl after irradiation using 436 nm light source.

Metal complexes of nickel (II), cobalt (III), zinc(II), copper(II) and cadmium(II) with ethylenediamine (en) were prepared by Varand *et al.*[24] The metal complexes precipitate from the aqueous solutions when high concentrations of the ethylenediamine were used. The syntheses of these complexes are illustrated in Scheme 3.2.

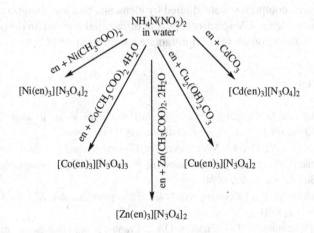

Scheme 3.2. Synthesis of metal complexes of dinitramide.

Table 3.4. Complex Salts of Dinitramide

Dinitramide	Melting point (°C)	Reference
$NaN_3O_4 . O\ \ O$	119–122	[1]
$AgN(NO_2)_2 . MeCN$	68–72	[1]
$[Ag(NH_3)_2]N(NO_2)_2$	58–64 (decomp)	[1]
$[Ag(py)_2]N(NO_2)_2$	68–69 (decomp)	[1]
$[Ag(O\ \ NH)_2)N_3O_4]$	125–126 (decomp)	[1]
$[Cu(NH_3)_4]\ [N(NO_2)_2]_2$	178–183 (decomp)	[1]
$[Cu(py)_4]\ [N(NO_2)_2]_2$	138–140 (decomp)	[1]
$[Ni(NH_3)_6]\ [N(NO_2)_2]_2$	149–155 (decomp)	[1]
$[Pd(NH_3)_4][N(NO_2)_2]_2$	168–174 (decomp)	[25]
$[Ag(NCCH_3)]N(NO_2)_2$	–	[25]
$[Ni(en)_3]\ [N(NO_2)_2]_2$	–	[24]
$[Co(en)_3]\ [N(NO_2)_2]_3$	–	[24]
$[Cu(en)_3]\ [N(NO_2)_2]_2$	–	[24]
$[Cd(en)_3]\ [N(NO_2)_2]_2$	–	[24]
$[Zn(en)_3]\ [N(NO_2)_2]_2$	–	[24]
$fac\text{-}Re(bpy)(CO)_3N_3O_4$	–	[23]
$[Hg(PhNH_2)_2]\ [N(NO_2)_2]_2$	145 (decomp)	[26]
$[HgPy_2]\ [N(NO_2)_2]_2$	176–178 (decomp)	[26]
$[Hg(Ph_3P)_2]\ [N(NO_2)_2]_2$	165–170	[26]
$[Hg(Me_2S)_2]\ [N(NO_2)_2]_2$	133–134 (decomp)	[26]
$Hg(N_3O_4)_2 . 2CH_2(COOEt)$	63–64 (decomp)	[26]

The prepared complexes were studied by elemental analysis, magnetic susceptibility measurements, UV spectroscopy and differential thermal analysis. A summary of the known complexes of dinitramide are given in Table 3.4.

References

1. OA Lukyanov, OV Anikin, VP Gorelik, *et al.,* Dinitramide and its salts 3. Metallic salts of dinitramide, *Russ Chem Bull* **43**(9): 1457–1461, 1994.
2. OA Lukyanov, VP Gorelik, VA Tartakovsky, Dinitramide and its salts 1. Synthesis of dinitramide salts by decyanoethylation of N,N-dinitro-β-aminopropionitrile, *Russ Chem Bull* **43**(1): 89–92, 1994.
3. VA Tartakovsky, OA Lukyanov, Synthesis of dinitroamide salts, *25th Int Annu Conf ICT,* **13**: 1–9, 1994.
4. VA Shlyapochnikov, NO Cherskaya, OA Lukyanov, *et al.,* Dinitramide and its salts 8. Synthesis, spectra and the structure of mercury(II) dinitramidate, *Russ Chem Bull* **45**(2): 430–432, 1996.

5. TM Klapoetke, P Mayer, A Schulz, *et al.,* 1,5-diamino-4-methyltetrazolium dinitramide, *J Am Chem Soc* **127**(7): 2032–2033, 2005.

6. (a) G Drake, T Hawkins, A Brand, *et al.,* Energetic low-melting salts of simple heterocycles, *Prop Expl Pyro* **28**(4): 174–180, 2003. (b) AJ Brand, GW Drake, Energetic hydrazinium salts, *US Patent 6218577 B1,* 2001.

7. MA Hiskey, MM Stineciper, JE Brown, Synthesis and initial characterisation of some energetic salts of 3,3-dinitroazetidine, *J Energ Mater,* **11:** 157–166, 1993.

8. VP Sinditskii, AE Fogelzang, AI Levshenkov, *et al.,* Combustion behaviour of dinitramide salts, *AIAA Paper 98-0808.*

9. D Chavez, MA Hiskey, 1,2,4,5-Tetrazine based energetic materials, *J Energ Mater,* **17:** 357–377, 1999.

10. OA Lukyanov, AR Agevnin, AA Leichenko, *et al.,* Dinitramide and its salts 6. Dinitramide salts derived from ammonium bases, *Russ Chem Bull* **44**(1): 108–112, 1995.

11. A Martin, AA Pinkerton, RD Gilardi, *et al.,* Energetic materials: The preparation and structural characterisation of three biguanidinium dinitramides, *Acta Cryst B* **53:** 504–512, 1997.

12. Lee Mia Nagao, PhD Thesis, Faculty of the Graduate School, Yale University, May 1998.

13. RJ Schmitt, JC Bottaro, PE Penwell, SRI International Final Report No. *AD-A 263271,* 1993.

14. RD Gilardi, RJ Butcher, A new class of flexible energetic salts, Part 5. The structures of two hexammonium polymorphs and the ethane-1,2-diammonium salts of dinitramide, *J Chem Cryst* **28**(9): 673–681, 1998.

15. T Hawkins, L Hall, K Tollison, *et al.,* Synthesis and characterization of energetic 1,2-Bis(oxyamino)ethane salts, *Prop Expl Pyro* **31**(3): 194–206, 2006.

16. R Tanbug, K Krischbaum, AA Pinkerton, Energetic materials: The preparation and structural characterization of melaminium dinitramide and melaminium nitrate, *J Chem Cryst* **29**(1): 45–55, 1999.

17. K Tollison, G Drake, T Hawkins, *et al.,* The synthesis and characterisation of methylene bisoxyamine $CH_2(-O-NH_2)_2$ salts. *J Energ Mat* **19**(4): 277–303, 2001.

18. H Ostmark, U Bemm, H Bergman, *et al.,* N-guanylurea-dinitramide: A new energetic material with low sensitivity for propellants and explosives applications. *Therm Chim Acta* **384:** 253–259, 2002.

19. H Boniuk, I Cieslowska-Glinska, M Syczewski, Synthesis and properties of salts of dinitroamide (DNA) with various amines, *31st Int Annu Conf ICT* **39:** 1–9, 2000.

20. HG Ang, W Fraenk, K Karaghisoff, *et al.,* Synthesis characterisation and crystal structures of various energetic urotropinium salts with azide, nitrate, dinitramide and azotetrazolate counter ions, *Z Anorg Allg Chem* **628:** 2901–2906, 2002.

21. N Wingborg, NV Latypov, Triaminoguanidine dinitramide, TAGDN: Synthesis and characterization, *Prop Expl Pyro* **28**(6): 314–318, 2003.

22. MA Petrie, JA Sheehy, JA Boatz, *et al.,* Novel high-energy density materials. Synthesis and characterisation of triazidocarbenium dinitramide, -perchlorate, and –tetrafluoroborate, *J Am Chem Soc* **119**: 8802–8808, 1997.

23. S Trammell, PA Goodson, BP Sullivan, Coordination chemistry and photoreactivity of the dinitramide ion, *Inorg Chem* **35**:1421–1422, 1996.

24. VL Varand, SV Larionov, NN Kundo, Metal complexes containing ethylenediamine and dinitramide anion, *Russ J Gen Chem* **69**(2): 263–264, 1999.

25. HG Ang, W Fraenk, K Karaghisoff, *et al.,* Synthesis characterisation and crystal structures of Cu, Ag, and Pd dinitramide salts, *Z Anorg Allg Chem* **628**: 2894–2900, 2002.

26. OA Lukyanov, OV Anikin, VA Tartakovsky, Dinitramide and its salts 9. Mercury(II) dinitramidate, a new reagent in chemistry of organomercury compounds, *Russ Chem Bull* **45**(2): 433–440, 1996.

Chapter 4

CHARACTERISATION OF ADN

The spectral, thermal and structural properties of ADN have been extensively investigated using a wide range of measurements. The results are presented in the following sections.

4.1. Spectral Methods

Spectral methods such as UV, IR, Raman and NMR spectroscopy were employed for the characterisation of ADN.

4.1.1. UV Spectroscopy

Bottaro et al. have characterised ADN using UV spectroscopy.[1] The UV spectrum showed absorption maxima at 212 nm and 284 nm with a molar extinction coefficient $E_{284} = 5207$ Lmol^{-1}cm^{-1}. The UV spectrum of ADN in water recorded in our laboratory is shown in Fig. 4.1. Absorption maxima at 208.8 nm due to high energy σ-σ^* transition and at 284.8 nm due to low energy n-π^* transition were observed.[2]

Ostmark et al.[3] showed that the UV-VIS spectrum of ADN has absorption maxima at 214 nm and 284 nm with a molar extinction coefficient of $E_{214} = 6680$ Lmol^{-1}cm^{-1} and $E_{284} = 5500$ Lmol^{-1}cm^{-1}. UV spectrum of ammonium dinitramide

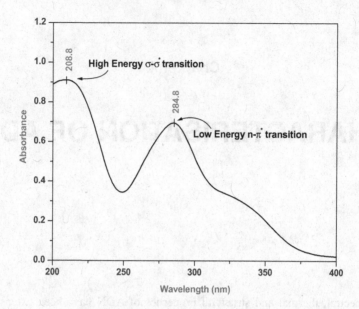

Fig. 4.1. UV spectrum of ADN in water.

in solution resembles the spectrum of the potassium salt of dinitramide. Absorption maximum at 285 nm were observed as a dominating band in UV.[4] The absorption maxima occurs at 223 nm, 284 nm and a shoulder at 335 nm with a molar extinction coefficient $E_{284} = 5640$ Lmol^{-1}cm^{-1} as reported by Lukyanov et al.[5] The electronic spectra of ADN showed absorption maxima at 218 nm, 285 nm and a shoulder at 335 nm with a molar extinction coefficient at $E_{285} = 5627$ Lmol^{-1}cm^{-1}. In 0.1 N KOH, the electronic spectra of ADN showed absorption maxima at 223 nm, 285 nm and a shoulder at 335 nm with a molar extinction coefficient of $E_{285} = 5640$ Lmol^{-1}cm^{-1}.[6]

4.1.2. *IR Spectroscopy*

Christe et al. have given the peak assignments for ADN in solid as well as liquid state, based on experimental as well as theoretical data.[7] The vibrational spectral frequencies which are calculated and observed for ADN are given. The authors have also calculated the IR spectral frequency for potassium and cesium dinitramide salts. The authors observed that the vibrational spectrum of dinitramide is greatly influenced by the type of cations and the physical state. The calculated geometries by HF/6-31G*, MP2/6-31+G*, NLDF/GGA/DZVPP levels of theory and the experimental geometries of ADN are given. Calculated and

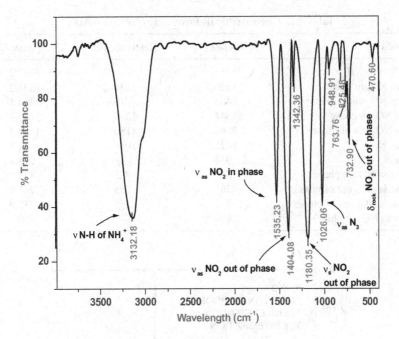

Fig. 4.2. FT-IR spectrum of ADN.

observed vibrational spectra of ADN, its solutions and their assignments in point group C_2 are given by the authors. A typical FT-IR spectrum of ADN in KBr pellet recorded in our laboratory[2] is shown in Fig. 4.2.

The characteristic IR frequencies for ADN as reported in the literature are summarised in Table 4.1.

4.1.3. Raman Spectroscopy

Fell *et al.* have recorded FT Raman spectra of neat ADN[8] employing near infrared laser radiation at a wavelength of 1.064 μm (9394.5 cm⁻¹) as the scattering source. The Raman shifts of the strongest bands for ADN are observed at 1335 cm⁻¹, 1043 cm⁻¹, 830 cm⁻¹, 493 cm⁻¹, and 163 cm⁻¹.

Ostmark *et al.* have carried out FT-Raman studies for the identification of ADN.[3] The authors conclude that FT-Raman measurements are simple to perform and give resolved sharp peaks which are easy to identify and quantify. The FT-Raman spectrum of ADN is shown in Fig. 4.3.

Christe *et al.*[7] have recorded Raman spectra of ADN in solid and liquid state. Shlyapochnikov *et al.* have measured the FT-Raman spectra of ADN in both solid

Table 4.1. Characteristic IR Frequencies (cm⁻¹) for ADN

Mode	Reference [7]	Reference [6]	Reference [2]
ν N-H of NH_4^+	3255	—	3132.2
ν as NO_2 in phase	1526	1530	1535.2
ν as NO_2 out of phase	1455	1410	1404.1
ν s NO_2 in phase	1344	1340	1342.4
ν s NO_2 out of phase	1181	1180	1180.4
ν as N_3	1025	1038	1026.1
ν s N_3	954	—	948.9
δ sciss NO_2 in phase	828	825	825.4
δ sciss NO_2 out of phase	761	755	763.8
δ rock NO_2 out of phase	727	735	732.9
δ wag NO_2 in phase	490	—	470.6

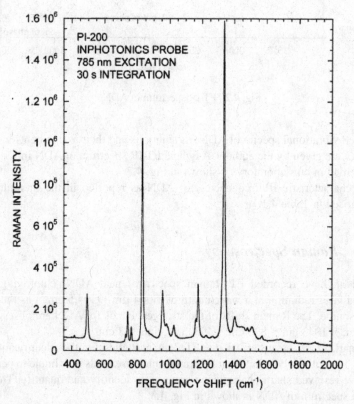

Fig. 4.3. FT-Raman spectrum of ADN. Reproduced with permission from Process Instruments Inc.

Table 4.2. FT-Raman Peak Assignments for ADN

Assignment	Reference [3]	Reference [8]	Reference [6]	Reference [7]
ν s NO_2 in phase	1338	1335	1335	1338
ν s NO_2 out of phase	1178	—	1174	1175
ν s N_3	958	1043	1025	1022
δ sciss NO_2 in phase	833	830	830	832
δ wag NO_2 in phase	495	493	493	492
δ sciss N_3	297	163	229	295

and liquid state.[6] A summary of the FT-Raman peaks for ADN as reported by various authors is given in Table 4.2.

4.1.4. *NMR Spectroscopy*

Shlyapochnikov *et al.* give an account on the structure and spectra of ammonium dinitramide and other salts.[9] The authors have prepared the isotopically substituted dinitramide salts and measured the ^{14}N and ^{15}N NMR spectra of these in solutions. With the use of *ab-initio* and other semiempirical calculations, the vibrational spectra of ADN have been measured with more accuracy. The authors have given a detailed account on the NMR spectra of the dinitramide salts.

Kaiser *et al.* characterised ADN by NMR spectroscopy.[10,11] Under high resolution conditions, they have recorded ^{14}N, ^{15}N and [17]O spectra of ADN and the spectra are shown in Figs. 4.4., 4.5. and 4.6. respectively. The ^{14}N NMR spectroscopy showed three signals at $\delta = -12.0$, -60.2 and -360.1 corresponding to the nitrogen of the nitro group, central nitrogen atom of dinitramide and the nitrogen atom of the ammonium ion respectively.

The measurement of ^{15}N spectra of ADN revealed three signals at $\delta = -12.2$, -60.8 and -360.1 corresponding to the nitrogen atom in the nitro group, the central nitrogen atom of the dinitramide and the nitrogen atom of the ammonium ion.

The ^{17}O NMR spectroscopy showed a signal at $\delta = 469.6$ which has been assigned to the oxygen atom in the nitro groups. The NMR peak values for ADN are summarised in Table 4.3.

4.1.5. *Mass Spectrometry*

The mass spectroscopic analysis of ADN showed major mass peaks at 30, 32, 44, 46 and 28, 30, 44, 46 for an electron impact potential of 20 eV and 70 eV

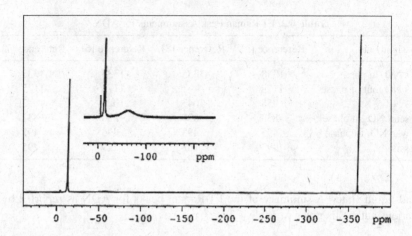

Fig. 4.4. [14]N NMR spectrum of ADN in D_2O. Reproduced with permission from Reference [11].

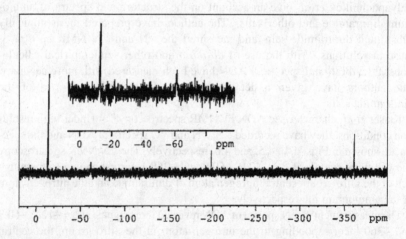

Fig. 4.5. ^{15}N NMR spectrum of ADN in D_2O. Reproduced with permission from Reference [11].

respectively.[3] The data indicates that ADN has poor stability and no mass fragments were observed above m/z 46 even at EI 20 eV. Vyazovkin *et al.* have obtained a complete mass spectrum and analysed the decomposition products in great detail.[12,13]

Doyle used a reverse-geometry tandem mass spectrometer to study the metastable and collision induced dissociation (CID) reactions of the dinitramide

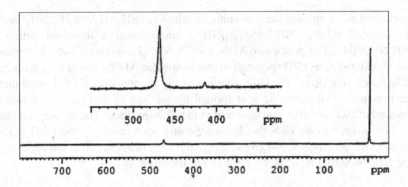

Fig. 4.6. ^{17}O NMR spectrum of ADN in D$_2$O. Reproduced with permission from Reference [11].

Table 4.3. NMR Peak Values for ADN

Nuclei	Chemical shift (ppm) Reference [10, 11].
^{14}N	−12.0 (N$\underline{N}$O$_2^-$)
	−60.2 (N$\underline{N}$O$_2^-$)
	−360.1 ($\underline{N}$H$_4^+$)
^{15}N	−12.2 (N$\underline{N}$O$_2^-$)
	−60.8 (N$\underline{N}$O$_2^-$)
	−360.1 ($\underline{N}$H$_4^+$)
^{17}O	496.6 (NN$\underline{O}_2^-$)

anion, the ammoniated form of ADN, protonated ADN and some ions which were formed by direct sputtering on the surface of ADN.[14] The negative ion mass spectrum of sputtered ADN showed a base peak at m/z 106 corresponding to the high abundance of dinitramide anion. Significant amounts of NO$_2^-$, oxygen radical anion and N$_2$O$_2^-$ were observed along with a peak at m/z 62, corresponding to the nitrate anion (m/z = 62). The nitrate ion might have resulted from the condensed-phase reactions during the sputtering process or the ion-molecule reactions in the high pressure region near the surface of the sample or the dissociation of the isomer of the dinitramide anion. The metastable dissociation spectrum of ADN showed the presence of NO$_2^-$, N$_2$O$_2^-$, NO$_3^-$ and N(NO$_2$)$_2^-$. The CID spectrum of the dinitramide anion showed the formation of O$^-$ and NO$^-$ in low abundance and high abundance of NO$_2^-$ and N$_2$O$^-$. But the relative abundance of the nitrate anion is low compared to that of the metastable spectrum. The positive-ion mass spectrum of ADN showed the presence of the base peak of sputtered ammonium ion apart from other species such as ammonia clusters, $NH_3NH_4^+$, $(NH_3)_2NH_4^+$ and ammoniated ADN, $(NH_4N(NO_2)_2)NH_4^+$, ammoniated ammonium nitrate, $(NH_4NO_3)NH_4^+$. The protonated ADN, $(NH_4N(NO_2)_2H^+$ is detected at a low relative abundance. The CID spectrum of the protonated ADN showed the presence of NO^+ and NO_2^+ ions. The CID spectrum of the ammoniated ADN showed that the presence of dinitramidic acid formed by the loss of $NH_3NH_4^+$ also small amounts of NO^+ and NO_2^+ are observed in the CID spectrum. The mass spectra in the high-mass regions of both the negative and positive-ion of sputtered ADN revealed the presence of $NH_4(N(NO_2)_2)_n[N(NO_2)_2]^-$ in the negative ion spectrum and $NH_4(N(NO_2)_2)_nNH_4^+$ in the positive ion spectrum.

4.2.   Chromatographic Methods

The experimental conditions and the results obtained on the characterisation of ADN by ion chromatography, HPLC and capillary electrophoresis are described.

4.2.1.   *Ion Chromatography and Capillary Electrophoresis*

Oehrle has carried out ion chromatographic analysis of ADN.[15] The method was developed to analyse as well as to find out the presence of anionic impurities such as nitrate, nitrite, sulfate etc. The presence of ADN was detected by UV by monitoring the wavelength at 214 nm. The disadvantage of the detection by UV was that sulfate ions cannot be detected as they don't absorb in the UV region. The author has used a 5 cm IC Pak-A IC column and a mixture of 3.0 mM lithium hydroxide and 20% MeCN as eluent. A typical ion chromatogram is shown in Fig. 4.7. The author has also done aging studies of ADN solutions exposed to light for different periods of time. Because of the high degree of sensitivity and selectivity of ion chromatography, the technique can be used to find out the purity as well as the amount of impurity in ADN.

Bunte *et al.* have analysed ADN, its precursor and byproducts using ion chromatographic technique.[16,17] The authors have identified with high precision the nitrate, nitrite and dinitramide anions using a suitable column and detector. The use of ANION R column with electrical conductivity detector and ANION R column with a UV detector failed to detect the presence of dinitramide anion. The possible reason could be the delocalisation of the dinitramide ion yielding a small difference in the conductivity and also because of high levels of eluent viz p-hydroxybenzoic acid used. The authors have overcome the problem by using a special anion exchange column DIONEX (IonPac 11 HS) coupled with a UV detector using sodium hydroxide as eluent. Ionic strength and flow of eluent have some effect on the elution time. The detection of dinitramide ion was done

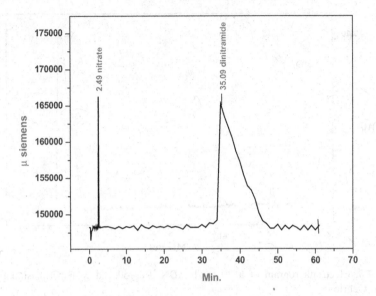

Fig. 4.7. Ion chromatogram of ADN.

at 285 nm and the nitrate ions at 214 nm. A concentration of 300 mM of NaOH was found to yield good results for the separation of nitrate and nitrite ions. The authors have extended their studies to detect the presence of ammonium nitrourethane, a precursor used in their study in the preparation of ADN. Santhosh *et al.* have characterised ADN by ion chromatography.[18] IC spectrum of ADN shown in Fig. 4.7 has a peak at 2.49 min corresponding to the nitrate ion and a broad peak at 35 min due to the dinitramide anion. No peaks corresponding to any other impurities were observed.

Oehrle has also characterised ADN by the use of capillary ion electrophoresis.[19] The author has used indirect and direct detection of dinitramide using a chromate and phosphate electrolyte respectively. The dinitramide anion was detected in the wavelength of 254 nm and 185 nm when the indirect and direct methods were used. Using a phosphate electrolyte, a very good separation of ADN and its degradation products were achieved. A 25 mM phosphate — 0.5 mM CIA-Pak OFM anion electrolyte and a direct UV detection at 185 nm were used in his studies for the analysis. With the help of UV detector, the impurities such as chloride, nitrate, nitrite and sulphate ions was detected precisely. The run can be completed in less than 5 min to 10 min and this method can be used to quantitatively determine the amount of ADN. The electropherogram of ADN is shown in Fig. 4.8. The peak at ~3.6 min is due to the chloride ion and the one at ~5 min is due to the dinitramide anion.

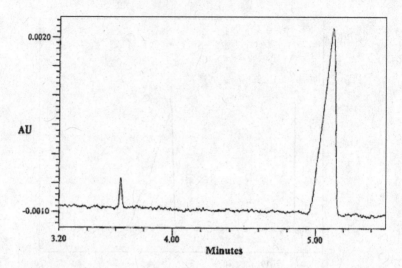

Fig. 4.8. Electropherogram of a 20 mg/L ADN. Reproduced with permission from Stuart Oehrle.

4.2.2.　HPLC Method

Holmgren *et al.* have developed a method for the characterisation of ADN by HPLC-UV.[20] The presence of ionic impurities is analysed on a porous graphitic column. The authors have used 99.5% water and 0.05% ammonia for dissolving the samples. The use of slightly basic solution is to prevent the decomposition of ADN. To lower the back pressure and reduce the analysis time, the columns were heated to 55°C. ADN is detected at the final stage of the analysis as a broad peak. A typical HPLC-UV chromatogram of ADN is shown in Fig. 4.9. The HPLC run was made using 0.05% ammonia solution as a mobile phase and the wavelength was set at 210 nm. The experiment was carried out at room temperature using a Hypercarb column.[2]

Fig. 4.9 showed a sharp peak at 3.03 min and a broad peak at 33.95 min which corresponds to the nitrate and dinitramide ions respectively. From the peak areas, the purity and the amount of impurities in ADN can be calculated quantitatively.

4.3.　Thermal Methods

Useful information such as melting point, heat of decomposition, phase transition, and weight loss were derived from TGA, DSC and DTA. The results on the thermal characterisation of ADN by the above methods are described.

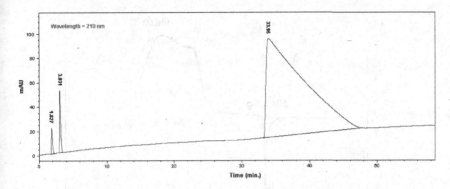

Fig. 4.9. HPLC-UV chromatogram of ADN.

4.3.1. DSC Method

Vyazovkin *et al.* have extensively studied the thermal decomposition of ADN by various thermoanalytical methods.[12,13] The DSC analysis showed a melting of ADN at 92°C followed by exothermic decomposition in the temperature range 130°C–230°C. The analysis of decomposition products, mechanism for condensed phase ADN decomposition and the reaction kinetics were performed and the results were summarised in their paper.

Thermal analysis of ADN was studied by Lobbecke *et al.*[11,21] The DSC studies indicate that ADN decomposes with an onset of 120.65°C ± 0.77°C after the melting at 90.37°C ± 0.5°C. The heat of decomposition was measured as 234.06 kJ/mol ± 5.5 kJ/mol. The decomposition product of ADN is identified as AN, whose endothermic decomposition can be clearly seen as an endotherm after the exothermic decomposition of ADN. The presence of AN in ADN has been studied by the authors, which showed that AN has significant influence on the melting behaviour of ADN. A typical DSC spectrum of ADN is shown in Fig. 4.10.

Langlet *et al.* have studied the thermal stability of ADN by DSC.[22] The activation energy and the frequency factor were calculated and found to be 158 kJ/mole and 15.8 s[-1] respectively. A comparison of the activation parameters with RDX revealed that ADN has a smaller activation energy and frequency factor. Bottaro *et al.* have characterised ADN by DSC.[1] The DSC results showed the melting of ADN at 92°C followed by an exothermic decomposition onset at 130°C. The DSC melting point of ADN is reported as 93.5°C.[3]

4.3.2. Thermogravimetric Analysis (TGA)

Ziru *et al.* have given an account on the thermal behaviour of ADN by TGA.[23] The authors have investigated the thermal decomposition characteristics, stability

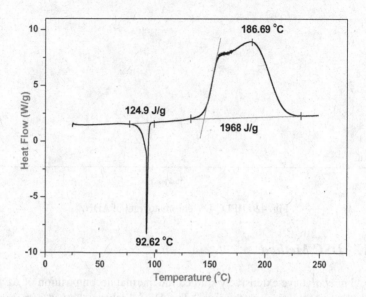

Fig. 4.10. DSC spectrum of ADN at a heating rate of 5°C/min.

and the degradation of ADN by PDSC and TG. The authors have studied the binary eutectic systems of ADN and AN. The melting points of the eutectic mixtures vary according to the amount of AN present. The thermal decomposition of ADN at high pressure and the influence of AN on ADN decomposition were studied in detail. They have derived the kinetics and mechanism for thermal decomposition of ADN, the decomposition mechanism and the preliminary analysis of the relationship between thermal decomposition and combustion of ADN. The results indicate that when AN was added to ADN, the multiple peaks observed were turned to a single peak at high pressure. The addition of AN forms an eutectic with an eutectic temperature of 60°C corresponding to 30% of AN. The thermal decomposition of ADN is accelerated in the initial stage and in the latter stage, it is inhibited by the addition of AN. At high pressure, ADN accelerates the decomposition process of AN. A typical TG-DTA trace of ADN is shown in Fig. 4.11.

The TG trace of ADN showed a single stage decomposition with 100% mass loss at 220°C. DTA trace of ADN showed a melting endotherm at 92°C followed by exothermic decomposition in the temperature range of 150°C–210°C with a peak maximum at 172°C. The DTA trace also shows an endotherm at 214°C, which is due to the endothermic decomposition of ammonium nitrate *in situ* formed during the decomposition of ADN.

By applying HiRes TGA, Lobbecke *et al.* have identified a two stage decomposition for ADN.[11,21] With the help of FT-IR spectroscopy combined with EGA,

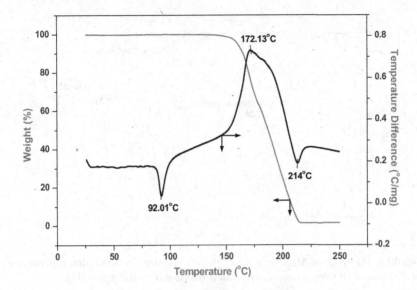

Fig. 4.11. TG-DTA trace of ADN at a heating rate of 5°C/min.

the gaseous products of ADN are identified. Decomposition fragments N_2O, NH_4NO_3, H_2O, NO_2, NH_3 and NO were detected. Based on the HiRes TGA and EGA data, ADN is decomposed by forming N_2O and NH_4NO_3 in the first stage (corresponding to 35% mass loss). The decomposition of formed NH_4NO_3 contributes to the other mass fragments. The TGA spectra of ADN showing a single and two step decomposition are shown in Fig. 4.12. The authors have also investigated the slow thermal decomposition of ADN by TM, isothermal TGA, isothermal EGA and MDSC experiments. The measurements show that the thermal stability of ADN is poor after its melting.

The thermogravimetric analysis by Vyazovkin *et al.* showed that the mass loss of ADN occurred in the same range as that of DSC experiments.[12,13] The decomposition of ADN followed a single step. By means of TG-MS experiments, they have identified NH_3, H_2O, NO, N_2O, NO_2, HONO and HNO_3 fragments during the thermal decomposition.

4.3.3. *Microcalorimetry*

Langlet *et al.*[22] have measured the stability of ADN at 65°C, 80°C and 90°C by microcalorimetry. Their observation indicates that impure and moist ADN show significant thermal activity, while the molten and pure ADN show stable thermal activity. Wingborg *et al.* studied the thermal stability of ADN by measuring the

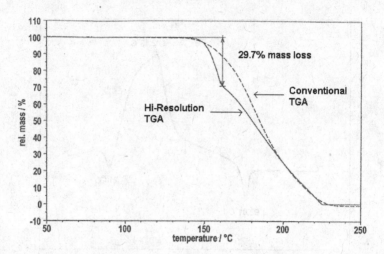

Fig. 4.12. TG traces of ADN showing single and two step decomposition patterns at a heating rate of 10°C/min. Reproduced with permission from Reference [11].

heat flow from 0.5 g of sample using a microcalorimeter.[24] The authors have measured the thermal stability of ADN between 55°C and 80°C. The results obtained from their studies are shown in Table 4.4.

The microcalorimetric study indicated a high thermal activity at 60°C and 65°C and is more stable at other temperatures. This behaviour is only obtained for samples that are very dry with a very less amount of moisture (0.01–0.05). To prevent the anomalous decomposition, ADN should be stored below 60°C.

4.4. Other Methods of Characterisation

A number of experimental and theoretical methods were used for the characterisation of ADN. The methods and the results from the study are described in the following sections.

Table 4.4. Total Energy Evolved for ADN as Measured by Microcalorimetry

	55°C	60°C	65°C	70°C	75°C	80°C
Heat Evolved (J/g)	0.0	57.6	25.6	−0.7	1.5	2.3
	0.1	75.8	23.7	—	2.7	2.7
	—	70.9	—	—	—	—

4.4.1. *X-ray Diffraction*

The crystal structure of ammonium dinitramide was determined by Gilardi *et al.*[25] ADN shows extensive hydrogen bonding between the four ammonium protons with the four oxygen atoms in the dinitramide anion. The crystal data, intensity data and structural refinement of ADN are given by the authors. ADN has inequivalent N-N and N-O bond distances and the nitro groups are twisted from the NNN bond angle. The metrical parameters predicted by theoretical means are in good agreement with the experimental results. The structural parameters for ADN from X-ray data are shown in Table 4.5.

The X-ray crystal structure data for ADN by Gilardi *et al.* are given in Table 4.6.

Ostmark *et al.*[3] have measured the bulk density of ADN using powder X-ray diffraction and single crystal X-ray diffraction at 293 K and found 1.8139 g/cm^3 and 1.8183 g/cm^3 respectively.

Table 4.5. Structural Parameters from X-ray Results for ADN

	Bond angles (deg.)		Bond lengths (A°)	
	N2-N1-N3	113.2	N1-N2	1.376
	N1-N2-O2A	113.03	N1-N3	1.359
	N1-N3-O3A	112.4	N2-O2A	1.236
	N1-N2-O2B	123.38	N3-O3A	1.252
	N1-N3-O3B	125.14	N2-O2B	1.227
	O2A-N2-O2B	123.35	N3-O3B	1.223
	O3A-N3-O3B	122.18		

Table 4.6. X-ray Crystal Data for Ammonium Dinitramide

Property	
System	Monoclinic
a	6.914 A°
b	11.787 A°
c	5.614 A°
β	100.4°
Density	1.831 g/cm^3
Crystal shape	Prism
Crystal color	Colorless

4.4.2. *T-Jump/FT-IR Spectroscopy*

Brill *et al.* have studied the decomposition of ADN[26] near the burning surface temperature by T-jump/FTIR of thin films heated at $2\,000°C/sec$ to $260°C$. The experimental setup is shown in Fig. 4.13. The authors have detected HNO_3, NH_3 and N_2O in equal amounts in the first stage and minor quantities of NO_2, AN and H_2O at the later stage. The possible reactions for the gaseous products released during high rate pyrolysis were also described.

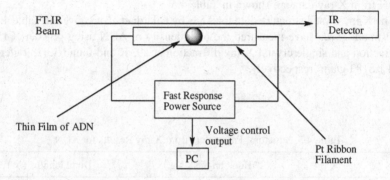

Fig. 4.13. T-Jump FT-IR experimental setup. Reproduced with permission from Begell House, Inc. Reference [27].

4.4.3. *Gas phase Study of Dinitramide*

Schmitt *et al.* have studied the gas phase chemistry of dinitramide anions.[28] The basicity, electron binding energy and collision induced dissociation were characterised for the anion. Their studies showed that the anion is extremely stable, unreactive in the gas phase, is a poor nucleophile and does not undergo electron transfer with moderately high electron affinity groups. The protonated form of the anion is extremely acidic in the gas phase. The collision-induced dissociation of the dinitramide anion produced $N_2O_2^-$ as a major product at m/z 60 and trace amount of NO_2^- at m/z 46.

4.4.4. *Theoretical Methods*

The thermochemistry and the structure of hydrogen dinitramide[29] were given by Michels and Montgomery. *ab- initio* calculations were performed at the RHF and MP2 levels of theory using the standard 6-31G** and 6-311+G** basis sets.

The calculated minimum energy structure for the dinitramide anion is a vibrationally stable C2 structure with oxygen atoms above and below the plane of the three nitrogen atoms. The authors give a detailed account on the structure of the protonated form of the dinitramide anion by different resonance stabilised structures and their minimum energy states. The calculated heat of formation of hydrogen dinitramide at MP2/6-311+G**//MP2/6-31G** is 28.4 kcal/mol.

Politzer *et al.* have studied the structure and some decomposition reactions of the dinitramide anion using density functional (DF) methods.[30] The structure optimised by nonlocal DF calculations is in good agreement with the structure calculated by crystallographic means. The authors have also theoretically investigated the energetics of various decomposition reactions of the dinitramide anion on the basis of experimental data. Various N-N bond breaking studies were performed and the lowest energy requirement for the dissociation of $N(NO_2)_2^-$ into NNO_2^- and NO_2 is calculated as 49.8 kcal/mol.

References

1. JC Bottaro, PE Penwell, RJ Schmitt, 1,1,3,3-Tetraoxo-1,2,3-triazapropene anion, a new oxy anion of nitrogen: The dinitramide anion and its salts, *J Am Chem Soc* **119**: 9405–9410, 1997.

2. G Santhosh, Ang How Ghee, Unpublished results.

3. H Ostmark, U Bemm, A Langlet, *et al.*, The properties of ammonium dinitramide (ADN) Part 1. Basic properties and spectroscopic data, *J Energ Mater* **18**: 123–138, 2000.

4. VA Shlyapochnikov, GI Oleneva, NO Cherskaya, *et al.*, Molecular absorption spectra of dinitramide and its salts, *J Mol Struc* **348**: 103–106, 1995.

5. OA Lukyanov, VA Tartakovsky, Synthesis and characterisation of dinitramidic acid and its salts, in solid propellant chemistry — combustion and motor interior ballistics, *Prog Astro Aero* **185**: 207–220, 2000.

6. VA Shlyapochnikov, GI Oleneva, NO Cherskaya, *et al.*, Dinitramide and its salts 7. Spectra and structure of dinitramide salts, *Russ Chem Bull* **44**(8): 1449–1453, 1995.

7. KO Christe, WW Wilson, MA Petrie, *et al.*, The dinitramide anion, $N(NO_2)_2$, *Inorg Chem* **35**: 5068–5071, 1996.

8. NF Fell, JM Widder, SV Medlin, *et al.*, Fourier transform Raman spectroscopy of some energetic materials and propellant formulations II, *J Raman Spect* **27**: 97–104, 1996.

9. VA Shlyapochnikov, MA Tafipolsky, IV Tokmakov, *et al.*, On the structure and spectra of dinitramide salts, *J Mol Struc* **559**: 147–166, 2001.

10. M Kaiser, B Ditz, Characterisation of ADN and CL-20 by NMR spectroscopy, *Int Annu Conf ICT* **130**: 1–8, 1998.

11. S Lobbecke, M Kaiser, GA Chiganova, Thermal and chemical analysis, in U Teipel (ed.), *Energetic materials — Particle processing and characterisation*, Wiley-VCH Verlag GmbH, 367–401, 2005.

12. S Vyazovkin, CA Wight, Ammonium dinitramide: Kinetics and mechanism of thermal decomposition, *J Phys Chem A* **101:** 5653–5658, 1997.

13. S Vyazovkin, CA Wight, Isothermal and nonisothermal reaction kinetics in solids: in search of ways toward consensus, *J Phys Chem A* **101:** 8279–8284, 1997.

14. Robert J Doyle Jr, Sputtered ammonium dinitramide: Tandem mass spectrometry of a new ionic nitramine, *Org Mass Spectrom* **28:** 83–91, 1993.

15. SA Oehrle, Analysis of anionic impurities in ammonium dinitramide (ADN) Part I: Ion chromatographic analysis, *J Energ Mater* **14:** 27–36, 1996.

16. G Bunte, H Neumann, J Antes, *et al.*, Analysis of ADN, it's precursor and possible byproducts using ion chromatography, *33rd Int Annu Conf ICT* **117:** 1–14, 2002.

17. G Bunte, H. Neumann, J Antes, *et al.*, Analysis of ADN, Its precursor and possible by products using ion chromatography, *Prop Expl Pyro* **27:** 119–124, 2002.

18. G Santhosh, Ph D Thesis, Mahatma Gandhi University, Kerala, India, 2005.

19. SA Oehrle, Analysis of anionic impurities in ammonium dinitramide (ADN) Part II: Capillary ion electrophoresis, *J Energ Mater* **14:** 37–45, 1996.

20. E Holmgren, H Carlsson, P Goede, *et al.*, Quantitative and qualitative analysis of nitrate and nitrite in commercial ADN, *Int Annu Conf ICT* **108:** 1–6, 2003.

21. S Lobbecke, T Keicher, H Krause, *et al.*, The new energetic material ammonium dinitramide and its thermal decomposition, *Solid State Ionics* **101–103:** 945–951, 1997.

22. A Langlet, N Wingborg, H Ostmark, ADN: A new high performance oxidizer for solid propellants, in KK Kuo (ed.), *Challenges in Propellants and Combustion — 100 Years after Nobel*, 616–626, 1997.

23. Liu Ziru, Luo Yang, Yin Cuimei, *et al.*, Thermal behaviour of a new energetic material ammonium dinitramide, *Int Pyrotech Semin* 326–333, 1999.

24. N Wingborg, M van Zelst, Comparative study of the properties of ADN and HNF, *FOA-R-00-01423-310-SE*. 2000.

25. R Gilardi, J Flippen-Anderson, C George, *et al.*, A new class of flexible energetic salts: The crystal structures of the ammonium, lithium, potassium and cesium salts of dinitramide, *J Am Chem Soc* **119**(40): 9411–9416, 1997.

26. TB Brill, PJ Brush, DG Patil, Thermal decomposition of energetic materials 58. Chemistry of ammonium nitrate and ammonium dinitramide near the burning surface temperature, *Comb Flame* **92:** 178–186, 1993.

27. TB Brill, Surface pyrolysis phenomena and flame diagnostics by FTIR spectroscopy, in KK Kuo, TP Parr (eds.), *Non-Intrusive combustion diagnostics*, Begell House Inc, 191–208, 1994.

28. RJ Schmitt, M Krempp, VM Bierbaum, Gas phase chemistry of dinitramide and nitroacetylide ions, *Int J Mass Spectrom Ion Process* **117:** 621–632, 1992.

29. HH Michels, JA Montgomery Jr, On the structure and thermochemistry of hydrogen dinitramide, *J Phys Chem* **97:** 6602–6606, 1993.

30. P Politzer, JM Seminario, MC Concha, *et al.*, Density functional study of the structure and some decomposition reactions of the dinitramide anion $N(NO_2)_2^-$, *J Mol Struc (Theochem)* **287:** 235–240, 1993.

Chapter 5

CRYSTALLISATION AND PRILLING OF ADN

One important aspect during the formulation of propellants is the optimisation of particle size distribution and shape of the solid oxidizer. ADN crystallises with very high aspect ratios (10–100:1) with either needle or platelet shaped crystals. It also exhibits undesirable characteristics such as irregular shape, poor thermal stability, higher sensitivity and extreme hygroscopicity. The higher aspect ratio crystals are difficult to process because of high viscosity of the binder and oxidizer at low solid loading. The needle shaped crystals are in general more sensitive than smooth rounded particles. ADN crystals are suitably modified to overcome the inherent detrimental attributes and also to achieve a higher solid loading. The technologies currently being used to tailor the particle size of ADN are reviewed in this chapter.

The optical microscopic images of ADN resulting from the synthesis exhibiting an irregular crystal shape and poor morphology are shown in Fig. 5.1.

Significant progress on the crystal modification of ADN by prilling and crystallisation has been achieved by scientists worldwide. The poor attributes of ADN are either eliminated or minimised by prilling and emulsion crystallisation methods.[1] The above two processes have gained much importance in the recent years as the way to obtain spherical ADN particles of controlled size.

5.1. Prilling Process

For the energetic materials community, the process of prilling is widely known. It is practised for the large scale production of prills such as ammonium nitrate,

Fig. 5.1. Optical microscopic images of ADN (as synthesised).

urea and its admixtures. It is also widely used for the production of prills of many materials having a fairly acceptable melting point.[2,3] ADN prills are formed either by allowing the molten particles to fall by gravity onto a prilling tower with a counter current flow of cold inert gas or by melting and cooling of ADN in an inert medium under continuous stirring.

5.1.1. *Prilling Tower Method*

Highsmith *et al.* reported a process for making thermally stabilised prilled ADN particles.[4] The prilling tower used for the preparation of spherical ADN is shown in Fig. 5.2.

The process consists of feeding ADN through a hopper to the top of a tubular glass column having hot and cold zones. The hot and cold zones are controlled in such a way as to melt the ADN when it comes to the hot zone and to solidify the ADN when it comes to the cold zone. For effective solidification, a cold inert gas or liquid is passed in a counter current direction to the falling molten ADN. The solidified ADN is then collected at the bottom of the tower and screened to isolate different particle sizes.

The authors also highlight the use of thermal stabilisers such as urea, mono and poly hydrocarbon urea derivatives such as 1,1-dialkylurea or 1,3-dialkylurea to stabilise ADN. The authors prefer to use a processing aid such as fumed silica (Tulanox or Cab-O-Sil) to prevent the particle agglomeration. The ADN prills obtained by this process exhibit enhanced processability, higher bulk density, lower mix velocity, improved homogeneity of resultant products, fewer voids in prills and enhanced resistance to humidity. The ADN prills are spheroidally-shaped with multimodal particle size in the range of about 30 µm–40 µm and 110 µm–200 µm. The authors have characterised the prilled ADN using spectral

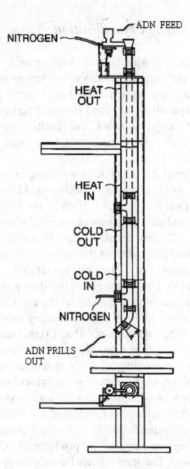

Fig. 5.2. Prilling tower used for making spherical ADN. Reproduced from Reference [4].

and thermal methods, and the safety data for the feed and the prilled ADN were compared. The results obtained indicate that the prills have good thermal, impact, and friction sensitivity. A paper highlighting the efforts put by the Swedish and the US governments towards realising and developing a manufacturing process for spherical ADN has also been published by the authors.[5] An account on the synthesis, prilling, application, compatibility and stability of ADN is also given in their paper. Highsmith *et al.* also disclosed a prilling process for few energetic materials other than ADN and AN.[6] Some of the meltable energetic materials used in their studies are trinitroazetidine (TNAZ), trinitrocresol, trinitroxylene, hexanitrodiphenylglycerol mononitrate, trinitronaphthalene, hydrazinium nitro-formate, and hydrazine nitrate.

5.1.2. Emulsion (Melt) Crystallisation

Emulsion crystallisation technique involves melting of a solid at slightly above its melting point in an inert liquid medium which serves as a continuous phase.[1] Under thorough stirring, the molten solid forms small and minute particles, which upon cooling, crystallises into hard spherical particles. Depending on the amount of mechanical energy supplied, the resulting particle sizes will vary. Improvements in the emulsion crystallisation have been achieved by suitably modifying the process.

In a patent by Wood *et al.* the apparatus for prilling an oxidising salt of ammonia is described.[7] The process involves melting of ADN with a stabiliser and injecting the molten ADN into an inert perfluorinated carrier liquid. The molten ADN and the carrier liquid then pass together in a turbulent flow through stationary vanes to disperse ADN into droplets. Then they enter a cooled conduit for solidification of ADN into prills without the formation of any agglomeration. The carrier liquid after the separation of the prills is recycled.

The authors highlight the disadvantages of the inert gas cooled prilling tower method. The disadvantages are: (1) cooling of the columns with 50 feet to 100 feet in height cannot be scaled down for a laboratory production; (2) control of the particle size and cooling rate is not possible; (3) the hazards involved in melting of large amounts of sample is high; (4) the corrosive nature of molten ADN to metals and; (5) the supercooling of ADN at temperatures below 70°C. The authors claim that the above are overcome by the present method. They have also given a detailed overview on the apparatus used for carrying out the experiment. The use of perfluorinated cyclic ethers identified as "FC-75" gave good results, as the boiling point of the same is 102°C with a pour point of −88°C. In another patent, the authors have given a process for producing ADN and AN prills by the method described above.[8] The method can be used to prepare stabilised ADN prills with a particle size in the range of 50 μm to 350 μm.

Chan *et al.* have studied the ADN prills obtained by a melt crystallisation process.[9] The prills were prepared by using a small laboratory scale apparatus. The process involves melting of ADN in a carrier fluid in which ADN is immiscible. The molten ADN in the carrier liquid is then subjected to mechanical stirring under a high shearing force. The temperature of the carrier fluid is maintained at 105°C, so as to melt the ADN. The molten ADN under a high mechanical agitation is then allowed to cool so as to form hardened spherical spheres. A thermal stabiliser can be conveniently added onto the molten ADN. The crystallised ADN is then separated from the carrier liquid and analysed systematically. The authors have subjected the prills to optical and scanning electron microscopic examination, vacuum thermal stability test, ageing studies and chemical characterisation by NMR and ion chromatographic techniques.

Pioneering work on the crystallisation of spherical ammonium dinitramide particles has been carried out by Teipel and coworkers.[10-12] Two stage emulsion crystallisation techniques for producing spherical ADN particles is described.[13-15] In the first stage, the molten ADN is dispersed in a continuous phase in which ADN is insoluble. The molten ADN is stirred rapidly by varying the amount of mechanical energy. In the second stage, crystallisation of the molten ADN droplets by lowering the temperature of the system is performed. The process flow diagram for the emulsion crystallisation is shown in Fig. 5.3.

The authors were able to get particle sizes in the range of 10 μm–600 μm. Paraffin oil as a carrier liquid was used and silicon dioxide with a particle size of 16 nm was used as a protective colloid to promote emulsification. The product analysis by spectral and thermal methods indicates that there is no significant difference between the synthesised and the crystallised ADN. An account on the crystallisation phenomena of molten ADN using a seed crystal is given by Teipel *et al.*[16] The formation of spherical particles of ADN by emulsion crystallisation incorporating additives, crystallisation initiators or stabilisers was patented by the author.[17] By varying the mechanical energy or by applying ultrasound, the process was modified to yield uniform spherical particles of ADN with a particle size ranging from few μm to few mm.

Santhosh *et al.* have prepared ADN prills by emulsion crystallisation.[18] The experimental setup for the process is shown in Fig. 5.4. The process involves the melting of ADN in paraffin oil maintained above the melting point of ADN. Paraffin oil is taken in a flat glass vessel (GV) and placed on a magnetic hotplate stirrer (HP). The temperature of the paraffin oil is controlled by a thermocouple

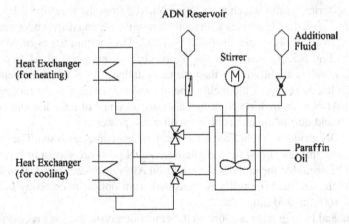

Fig. 5.3. Process diagram for the emulsion crystallisation of ADN. Reproduced with permission from U. Teipel.

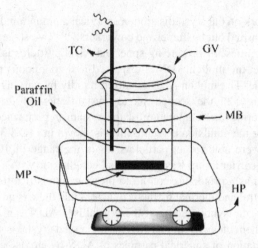

Fig. 5.4. Experimental setup for the emulsion crystallisation of ADN.

(TC) immersed in the oil. The stirring of the oil and molten ADN is achieved by placing a rectangular flat magnetic pellet (MP). The whole assembly is placed in a metallic bath (MB) to aid in cooling the glass assembly with a coolant. Once the required temperature of the paraffin oil i.e. 92°C–95°C is reached; the recrystallised ADN is slowly introduced into the paraffin oil under constant stirring.

ADN melts in paraffin oil and the stirring of the oil aids in forming molten ADN grains. After the complete formation of the molten grains, the whole assembly is slowly cooled to ambient temperature whereby the molten ADN crystallises to hard spherical grains which were later removed from the paraffin oil by filtration and then washed few times with methylene chloride and dried under vacuum.

The ADN prills were then analysed by DSC. The melting traces of ADN are shown in Fig. 5.5. As seen from the figure, the melting point of ADN prills is 91.89°C which is slightly lower than the recrystallised ADN which is melting at 92.44°C. The lowering of the melting point can be attributed to the thermal stability of ADN at its melting point and also the presence of trace amounts of the prilling liquid during the emulsion crystallisation process.

The ADN prills are subjected to optical microscopic analysis. The optical microscopic images of emulsion crystallised ADN prills are shown in Fig. 5.6.

It can be seen that the emulsion crystallised ADN prills are spherical in shape. The particle size of ADN prills as measured from optical microscopy is in the range of 100 μm–300 μm.

Heintz and Fuhr give an account on the emulsion crystallisation process for the preparation of spherical oxidizer particles.[19] An emulsion crystallisation process of ADN is explained whereby the molten ADN particles in an inert medium are

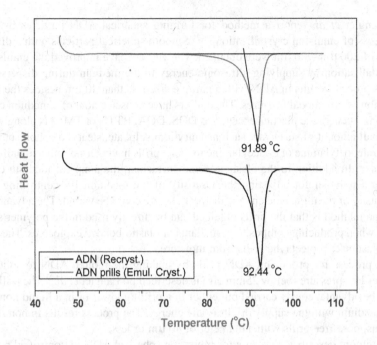

Fig. 5.5. DSC melting traces of recrystallised and emulsion crystallised ADN.

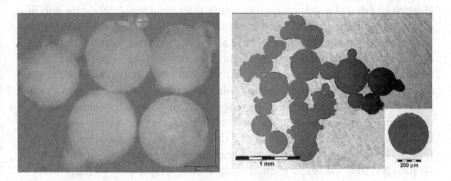

Fig. 5.6. Optical microscopic images of emulsion crystallised ADN.

allowed to cool slowly under agitation. The results indicate that spherical particles with a median size in the range of 50 μm–500 μm can be obtained. Because of the better control of the system and good particle size distribution, the authors suggest the use of emulsion crystallisation for ADN.

Langlet *et al.* report a method for forming spherical ADN particles by the process of emulsion crystallisation.[20-22] Smooth spherical particles with a diameter of 200 μm–700 μm were obtained. The authors have improved the emulsion crystallisation by supplying ultrasonic energy to the medium during dispersion. This process results in ADN with a particle size less than 40 μm besides the formation of polymodal mixtures. The authors have chosen a nonpolar medium consisting of energetic plasticisers such as DOS, DOA, BTTN or TMETN along with a small amount of surfactant such as laurylium sulphate, stearyl alcohol, sodium stearate, octylamine or dodecylamine to make prills having a smooth and uniform surface. In addition to the plasticiser and surfactant, the nonpolar medium contains a polymer for adjusting the viscosity of the medium. By controlling the intensity of the ultrasonic energy, the particle size could be varied. The advantage of the method is that the prills obtained can be directly used in the polymer system when producing composite propellants or plastic-bonded explosives. The use of plasticiser protects the prills from moisture.

A process for producing ADN prills by making the molten ADN to solidify under low pressure (below 25mm Hg) is described by Langlet *et al.*[23] The prilling can be advantageously carried out either in a prilling tower or in a liquid nonpolar medium with the supply of ultrasonic energy. The process results in homogeneous, pore-free prills with a diameter of 40 μm or less.

Guimont describes a process for preparing spherical ADN of controlled particle size with a limited melt processing time.[24] The process involves feeding the solid ADN at a controlled rate to a heating column to melt ADN and is then allowed to fall under gravity into a cooling fluid maintained below the solidification temperature of ADN. The fluid is vigorously stirred to form small uniform spherical ADN particles which solidify at a very short time. The solidified ADN is then removed from the inert fluid and dried further. The author also explains another process whereby a slurry of ADN and an inert fluid is fed in the hopper and allowed to pass through a heated column to melt the ADN and onto an inert fluid maintained below the melting temperature of ADN for solidification. The separation of ADN is done by the same way as stated in the former process. The author suggests the use of mineral oil, silicone oil and fluorocarbon oil as inert fluids. The majority of the particles obtained by the process are in the range of 10 μm–200 μm. The experimental setup for the preparation of ADN prills is shown in Fig. 5.7.

The application of emulsion crystallisation to form spherical particles of ADN is given by Heijden *et al.*[1] The crystallisation is carried from the molten ADN which is dispersed in an inert polar continuous phase. The authors explain the two stage crystallisation process whereby an emulsion of molten ADN as the disperse phase and paraffin oil as the continuous phase is prepared under agitation using a mechanical stirrer whose geometry and speed are selected for the system. The molten ADN grains are crystallised into hard spheres by giving mechanical

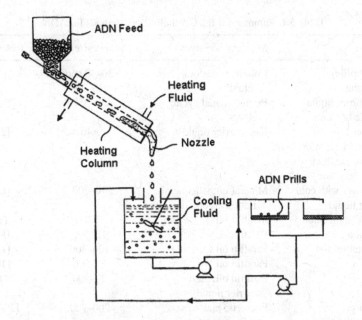

Fig. 5.7. Experimental setup for producing ADN prills. Reproduced from Reference [24].

energy, particle-particle interaction or supplying a seed crystal during the cooling of the reactor below the melting temperature of ADN. The crystallised ADN particles are then filtered and washed. The process produces spherical ADN in the particle size between 20 μm and 500 μm.

5.1.3. *Spray Prilling Process*

Heintz and Fuhr give an account on the spray crystallisation process for the preparation of spherical oxidizer particles.[19] The authors explains the production of spherical particles of AN by a spray crystallisation process and ADN prills by emulsion crystallisation.

A spray prilling process for ADN is recently reported by Johansson *et al.*[25] in which ADN is melted in a stainless steel vessel and was atomised by spraying through a nozzle under nitrogen pressure into liquid nitrogen. To achieve solidification, small quantities of Aerosil was added, the authors found that by maintaining low relative humidity during the prilling process the addition of Aerosil can be avoided. The particle size distribution varies while using nozzles of different diameter and it was not affected by varying the nitrogen pressure. The density and the thermal stability of the prills were tested and found to be good.

Table 5.1. Summary of the Crystallisation Methods for ADN

Method	Medium/Additives	Particle size (μm)	Reference
Glass prilling column	Cold inert gas or liquid	30–40 and 110–200	[4]
Inert carrier liquid cooled column	Perfluorinated cyclic ethers	50–350	[7,8]
Prill tower or crystallisation in liquid medium (low pressure)	Inert carrier liquid	<40	[23]
Prill tower with cold inert liquid	Mineral oil, silicone oil, fluorocarbon oil	10–200	[24]
Melt crystallisation	Paraffin oil	—	[9]
Emulsion crystallisation	Paraffin oil	10–600	[10–15]
	Paraffin oil	100–300	[18]
	Paraffin oil	50–500	[19]
	Paraffin oil, inert carrier liquid	20–500	[1]
Emulsion crystallisation (ultrasonic energy)	Energetic plasticisers, surfactants	200–700	[20–22]
Spray Prilling	Aerosil, liquid nitrogen	48–167 115–484	[25]

A summary of the crystallisation processes, experimental conditions and results for the preparation of ADN prills are given in Table 5.1.

5.2.   Thermal Initiation of ADN Crystals and Prills

Ramaswamy has studied the thermal initiation of ADN crystals and prills obtained by prilling tower and emulsion crystallisation methods using an environmental scanning electron microscope.[26] The initiation of the crystals and prills is achieved by either electron beam heating or a hot stage and the transformation was recorded in real time using a video camera. ADN crystals when subjected to electron beam heating showed microscopic "reaction sites" of the order of 0.01 μm during the start of the reaction and expands over the crystal surface at higher electron beam intensity leaving a hole by forming porous residue material. The crystallographic nature of the initiated crystals follows the same crystal habit on which they are formed. When the nitrogen atmosphere was changed to a moisture environment, the reaction is rapid where the rate of "reaction sites" has increased considerably.

The prills obtained from the prilling tower technique undergo a different type of initiation. The "reaction sites" are not homogenously distributed on the surface of the prills. About six times higher, electron beam intensity was required to initiate the prills. The reason is due to the presence of silica used during the prilling of ADN. The silica in the surface of the prills quenches the reaction sites. The presence of moisture increases the rate of decomposition for the prills and the presence of silica protects the prills from moisture.

The thermal initiation of prills obtained by emulsion crystallisation follows the same type of initiation as that of crystalline ADN. Along the cooling oil streamlines, the crystallisation occurs. The surface of prills is very rough and forms numerous defect sites. The use of hot stage in place of electron beam heating produced similar results for the crystals and prills. The author concludes that different manufacturing process produces crystals and prills with different microstructural characteristics and the initiation behaviour largely depends on the crystal structure. The presence of additives, impurity and stabilisers in the prills acts to reduce the threshold energy for the initiation.

References

1. AVD Heijden, J ter Horst, J Kendrick, *et al.,* Crystallisation, Ulrich Teipel (ed.), in *Energetic materials: Particle processing and characterisation*, Wiley-VCH publishers, 53–157, 2005.
2. FW Bennett, Prilling, *U.S. Patent 3997636*, 1976.
3. JK Bradley, Prilling, *U.S. Patent 3951638*, 1976.
4. TK Highsmith, CS Mcleod, RB Wardle, R Hendrickson, Thermally stabilised prilled ammonium dinitramide particles and process for making the same, *U.S. Patent 6136115 & WO Patent 99/01408.*
5. P Sjoberg, R Wardle, T Highsmith, A cooperative effort to develop manufacturing processes for spherical ADN, 2001 Insensitive munitions and energetic materials technology symposium, France, 466–470, 2001.
6. TK Highsmith, HE Johnston, Prilled energetic particles and process for making the same, *U.S. Patent 6610157,* 2003.
7. SE Wood, RA Weinhardt, Apparatus for prilling an oxidising salt of ammonia, *U.S. Patent 6135746,* 2000.
8. SE Wood, R.A. Weinhardt, Prilling by introduction of a molten liquid into a carrier liquid, *U.S. Patent 6074581,* 2000.
9. ML Chan, A Turner, L Merwin, ADN propellant technology, in Kenneth K Kuo (ed.), *Challenges in Propellants and Combustion: 100 Years after Nobel*, Begell House publishers, UK, 627–635, 1997.
10. U Teipel (ed.), *Energetic Materials : Particle processing and characterisation*, Wiley VCH publishers, 2005.

11. U Teipel, Product design of particulate energetic materials, *Mat Sci Technol* **22**(4): 414–421, 2006.

12. U Teipel, Particle design of energetic materials, *Cent Eur J Energ Mat* **2**(2): 55–69, 2005.

13. U Teipel, T Heintz, HH Krause, Crystallisation of spherical ammonium dinitramide (ADN) particles, *Prop Expl Pyro* **25**: 81–85, 2000.

14. U Teipel, T Heintz, On the crystallisation of ammonium dinitramide, *Int Annu Conf ICT* **49**: 1–12, 2001.

15. U Teipel, Particle technology: Design of particulate products and dispersed systems, *Chem Eng Technol* **27**(7): 751–756, 2004.

16. U Teipel, T Heintz, Surface energy and crystallisation phenomena of ammonium dinitramide, *Prop Expl Pyro* **30**(6): 404–411, 2005.

17. U Teipel, H Krause, K Leisinger, *et al.,* Production of uniformly shaped and sized particles of meltable propellants, explosives and oxidizing agents, *European Patent EP0953555*, 1999.

18. G Santhosh, Ang How Ghee, Unpublished results.

19. T Heintz, I Fuhr, Generation of spherical oxidiser particles by spray and emulsion crystallisation, *VDI-Berichte Nr. 1901*, 471–476, 2005.

20. A Langlet, M Johansson, Method of producing ADN prills, *WO Patent 99/21795*, 1999.

21. A Langlet, N Wingborg, H Ostmark, ADN: A new high performance oxidiser for solid propellants, in KK Kuo (ed.), *Challenges in Propellants and Combustion: 100 Years after Nobel*, Begell House publishers, UK, 616–626, 1997.

22. P Goh, M Johansson, Prilling of ammonium dinitramide (ADN), FOI-R—639—SE, 2002.

23. A Langlet, M Johansson, Method of producing prills of ammonium dinitramide (ADN), *WO Patent 99/21793,* 1999.

24. J Guimont, Process for preparing spherical energetic compounds, *U.S. Patent 5801453,* 1998.

25. M Johansson, John de Flon, A Petterson, *et al.,* Spray prilling of ADN and testing of ADN-based solid propellants, *3rd Int Conf on Green Propellants for Space Propulsion*, 2006.

26. AL Ramaswamy, Study of the thermal initiation of ammonium dinitramide (ADN) crystals and prills, *J Energ Mater* **18**: 39–60, 2000.

Chapter 6

COATING AND MICROENCAPSULATION OF ADN

The process of surrounding or enveloping a material within another material or a substance on a very small scale is known as microencapsulation. A semi or nonpermeable layer is formed by using a suitable coating material that is resistant to a particular environment. The particle size of core material used can be from $1\,\mu m$–$500\,\mu m$, surrounded by a layer of encapsulating material. The core material is spherically shaped and the encapsulating material is normally of natural or synthetic polymers.

ADN is highly hygroscopic and readily soluble in water. The magnitude of hygroscopicity is higher than that of AN. When used in propellant formulation, the hygroscopicity will have strong detrimental effects on combustion, storage and shelf life of the propellant. Hence, it is of utmost importance to prevent the ADN crystals and prills from the interaction of moisture. The coated ADN will have reduced sensitivity, improved moisture resistance and chemical compatibility. The processes evolved for the coating and microencapsulation of ADN are reviewed.

6.1. Microencapsulation via Polycondensation or Addition

The microencapsulation of particulate materials is reviewed by Teipel et al.[1] and the various processing approaches are briefly explained. The mechanical/physical

microencapsulation employs a gaseous carrier fluid and the chemical microencapsulation uses a liquid carrier fluid for producing encapsulated materials. The
authors highlight three processes for the encapsulation viz. formation of a coating via polycondensation or addition, encapsulation via *in situ* polymerisation
and coating under supercritical conditions. The polycondensation or addition
involves a reaction of two different monomers at the interface resulting in the
formation of a polymer film. *In situ* polymerisation involves the formation of a
polymer film onto a core material by a catalyst activation of a monomer or pre-
condensate. The supercritical coating involves the use of supercritical CO_2
($scCO_2$) whereby CO_2 flows through a thermally stabilised chamber containing
the coating material. The CO_2 expands into the fluidised bed reactor leaving a thin
coat onto the core material.

The coating of ADN is achieved by using either ethyl cellulose with ethoxy
concentrations of 48%–49.5% or cellulose acetate butyrate with an acetic acid
concentration of 55%–60%.[1] The coating of ADN prills were accomplished
by cooling a solution of ethyl cellulose in cyclohexane or cellulose acetate
butyrate in a nonpolar solvent followed by separation and drying of the
coated prills.

A microencapsulation process for ADN was described by Heintz *et al.* using
suitable coating materials and solvents.[2] The microencapsulation of ADN was
achieved by coacervation by cooling a solution of ethyl cellulose in cyclohexane.
A suitable protective colloid is also added to prevent the particle-particle agglomeration. Ethyl cellulose with ethoxyl content of 48%–49.5% was used in their
studies. The authors have characterised the encapsulated prilled ADN particles
for the coating efficiency.

The production of microencapsulated ADN using wax based coating materials
is given by Teipel *et al.*[3] The ADN particles are coated in a two step process consisting of coating the ADN particles under the absence of moisture using a paraffin wax or poly ethylene glycol wax and the polymerisation of wax coated ADN
under aqueous conditions using urea-formaldehyde, melamine-formaldehyde or
urea-melamine-formaldehyde prepolymer amino resins.

6.2.　Coating of ADN using Moving Bulk Solid Technique

The coating of spherical ADN particles is described by Heintz *et al.* using a
moved bulk solid technique with a pelletiser plate of 400 mm diameter operating
under an angle of 30°C to 75°C with a variable control for the speed regulation.[4]
The coating materials are dissolved in heptane or cyclohexane and sprayed onto
the ADN particles and allowed to evaporate, whereby a thin layer of coating is
achieved. The chosen coating materials are polymer I, polymer II, polymer II

with zinc particles, polymer II with zinc and aluminium particles, polymer III with aluminium particles and glycidyl azide polymer (GAP). The authors have characterised the coated ADN particles using light microscopy, scanning electron microscope and heat flow calorimetry. The results show that polymer II and GAP showed better compatibility for ADN, while polymer I, polymer II and polymer II with aluminium particles offered a free flowing powder with less agglomeration.

6.3. Supercritical Carbon Dioxide aided Coating of ADN

The supercritical carbon dioxide coating of ADN is given by Nauflett and Farncomb.[5] The authors highlight the use of nontoxic, noncorrosive and low cost CO_2 as an environmentally acceptable solvent for polymer based coating on ADN. In their studies, ADN was coated with 1% Viton. The coating method involves the addition of the polymer and ADN to a pressure vessel and increasing the temperature and pressure of CO_2. The temperature is then decreased and the CO_2 pressure is lowered. The weight percentage of the coating material onto ADN is determined by finding the difference in weight of coated and uncoated samples. The process generated virtually no hazardous waste. The coated ADN has low friction and electrostatic discharge.

A summary of the coating process used for ADN is given in Table 6.1.

Table 6.1. Coating Process and Materials used for ADN

Coating method	Coating material	Reference
Coacervation	Ethyl cellulose with ethoxy content 48%–49.5%.	[2]
Wax coating followed by *in situ* polymerisation	Paraffin wax, polyethylene glycol wax, urea formaldehyde, melamine formaldehyde or urea-melamine-formaldehyde prepolymer amino resins	[3]
Poly condensation or addition, *in situ* polymerisation under supercritical conditions	Cellulose acetate butyrate with Ac_2O content 55%–60%, cellulose esters	[1]
Super critical CO_2	Polymer based, 1% Viton	[5]
Moved bulk solid technique using a pelletiser	Different polymers, glycidyl azide polymer	[4]

References

1. U Teipel, T Heintz, H Krober, Microencapsulation of particulate materials, *Powder Handling and Processing* **13**(3): 283–288, 2001.
2. T Heintz, U Teipel, Coating of particulate energetic materials, *31st Int Annu Conf ICT* **120**: 1–12, 2000.
3. U Teipel, T Heintz, Production of microencapsulated moisture sensitive propellants, explosives and oxidising agents comprises wax coating raw material particles and encapsulating with amine resin, *European Patent DE 19923202, 2000.*
4. T Heintz, K Leisinger, H Pontius, Coating of spherical ADN particles, *37th Int Annu Conf ICT* **150**: 1–12, 2006.
5. GW Nauflett, RE Farncomb, Coating of PEP ingredients using supercritical carbon dioxide, CPIA publication, 708 (*JANNAF 30th Prop Dev and Charac. Subcomm. Meeting*), 13–19, 2002.

Chapter 7

STABILITY AND THERMAL STABILISATION OF ADN

The important characteristic of new propellants based on ADN is their chemical stability. The practical applicability of ADN as an oxidizer in solid propellants is still not fully realised because of many issues related to the thermal stability, compatibility, hygroscopicity, reactivity, crystal morphology and cost. ADN shows a slow and continuous decomposition at temperatures 60°C–65°C and is autocatalysed by the products formed during the thermal decomposition. Compositions containing ADN tend to decompose slowly when aged at temperatures above ambient. Suitable stabilisers are needed to arrest/inhibit the decomposition of ADN and also to improve the thermal stability of ADN. The studies on the stability of ADN have gained a great deal of interest and importance in the recent years. The thermal stability, types of stabilisers and its effectiveness along with the results obtained on the stabilisation of ADN are reviewed in this chapter.

7.1. Stability of ADN

Studies on the synthesis of ADN have been reported since its first synthesis.[1–2] Issues such as hygroscopicity, thermal stability and compatibility with propellant ingredients are being studied worldwide.[3–4] The stability of ADN has been studied by many authors. ADN is stable to bases and decomposes in the presence of acids. Pavlov et al. observed unusual regularities for the decomposition of ADN in the solid phase.[5] The solid phase reaction is 10–10^4 times faster than that in the melt. The decomposition is influenced by the formation of eutectics with the decomposition products and also by water. The maximum reaction rate is observed at the melting point of ADN and its eutectic mixtures with NH_4NO_3. Manelis *et al.* found that the decomposition rate of dried ADN with a water content of 0.05% is about one thousand times faster that of ADN with 0.5% moisture.[6] Acids and water show greater influence on the decomposition of ADN. The stability order of ammonium salts are $NO_3^- > N(NO_2)_2^- > NO_2^- \gg alkyl-N(NO_2)_2$. Thus ADN is more stable than NH_4NO_2 and alkyl dinitramides. Pak *et al.* studied the chemical stability of ADN in molten and solution state as well as with various admixtures.[7] Their study indicates that water is chemically inert to ADN and the rate of decomposition of aqueous ADN is two orders of magnitude higher than that in crystalline state. In organic solvents, the stability is considerably greater than that of water and is affected by the dielectric constant. The formation of low temperature eutectics with various chemically inert salts acts to accelerate the thermal decomposition of ADN. The thermal stability of ADN is dependant on the purity.

7.1.1. Reactions Steps for the Decomposition of ADN

The primary decomposition step is the rearrangement of ADN to ammonia and dinitramidic acid. The dinitramidic acid being unstable dissociates into HNO_3 and N_2O. The ammonia and HNO_3 formed combine together to form ammonium nitrate, which can decompose at higher temperatures forming N_2O and H_2O. At higher temperatures (above 160°C), the decomposition products are similar to those of AN, the dinitramidic acid decomposes via N-N bond scission leading to the formation of NO_2. The reaction steps shown in Scheme 7.1 are considered for the decomposition of ADN.

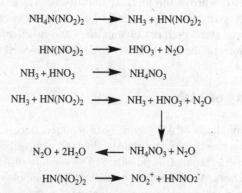

$$NH_4N(NO_2)_2 \longrightarrow NH_3 + HN(NO_2)_2$$

$$HN(NO_2)_2 \longrightarrow HNO_3 + N_2O$$

$$NH_3 + HNO_3 \longrightarrow NH_4NO_3$$

$$NH_3 + HN(NO_2)_2 \longrightarrow NH_3 + HNO_3 + N_2O$$

$$N_2O + 2H_2O \longleftarrow NH_4NO_3 + N_2O$$

$$HN(NO_2)_2 \longrightarrow NO_2^+ + HNNO2^-$$

Scheme 7.1. Reaction steps for the decomposition of ADN.

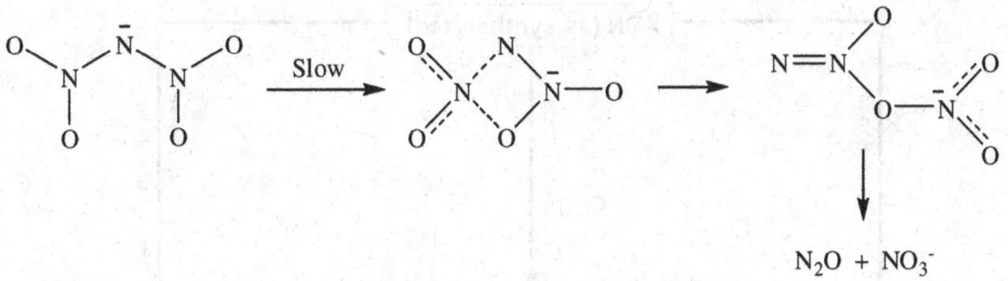

Scheme 7.2. Mechanism of formation of nitrate anion.

Ammonium nitrate is the primary decomposition product during the thermal decomposition of ADN. The formation of nitrate anion from the dinitramide anion is illustrated in Scheme 7.2.

7.2. Stabilisation of ADN

ADN can be stabilised either in solid or liquid phase using suitable stabilisers. The solid phase stabilisation primarily involves a thorough mixing of ADN with stabilisers until all the stabiliser particles are uniformly distributed in ADN. The liquid phase stabilisation involves either adding the stabiliser onto the molten ADN and allowing it to cool to ambient or by dissolving ADN and the stabiliser in a solvent and evaporating the solvent. The stabiliser can also be conveniently introduced during the process of emulsion crystallisation of ADN. A process flow sheet for the stabilisation is shown in Scheme 7.3.

First paper on the stabilisation of ADN appeared in the year 1992 by Russell *et al.*[8] They have investigated the polymorphic phases as a function of temperature, eutectic phase diagram of ADN/AN, thermal ageing and stabilisation of ADN. They have monitored the evolution of N_2O to study the ADN → AN rearrangement. *In situ* isothermal decomposition at 75°C was performed for ADN with stabilisers selected from 1) free radical scavenger → butylated hydroxytoluene (BHT), D-galactose (GAL) and 2,2-diphenyl-1-picrylhydrazyl (DPPH); (2) combined free radical scavenger → m-nitroaniline (MNA) and 2-nitrodiphenylamine (2-NDPA); (3) base → diphenylamine (DPA) and NH_3.

Stabilisers which are free radical scavengers showed no effect on the stabilisation. Retardation of decomposition of ADN was observed with combined free radical scavenger and bases. The time for initial N_2O evolution is observed at 15 min for the free radical scavengers, at eight hours for MNA and four hours for 2-NDPA samples. The authors have chosen 2-NDPA as a stabiliser at 1% level and systematically studied the evolution of N_2O by FT-IR spectroscopy. When

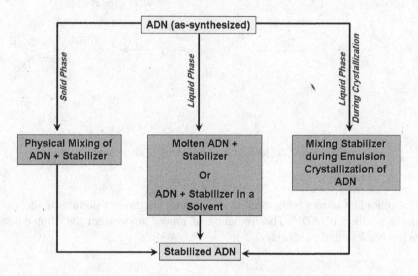

Scheme 7.3. Process flow diagram for the stabilisation of ADN.

aged at 75°C, no N_2O is observed for the initial six hours. Once the stabiliser concentration is depleted i.e. after six hours, the rate of formation of N_2O is equal to the rate of decomposition of neat ADN. The authors have given a mechanistic description of the low temperature decomposition based on their experimental data. They have described a two step process for the formation of N_2O from ADN, the initial step is the formation of dinitramidic acid and a unimolecular rearrangement of the latter to N_2O and HNO_3. Based on the calculations, the lowest energy mechanism shown in Scheme 7.4 is identified to be the decomposition pathway of ADN.

The authors have determined the rate determining step for the formation of NH_4NO_3 from NH_3 and HNO_3 based on their experimental data and the conversion of ADN → AN is considered very important in evaluating the thermal stability of ADN.

The thermal decomposition and stabilisation of ADN was studied by Lobbecke et al.[9] using five different stabilisers viz. magnesia (MgO), sodium metaborate ($NaBO_2$), hexamethylenetetramine (hexamine), 2-nitrodiphenylamine and methyldiphenylurea (akardit II). The stability and the decomposition behaviour

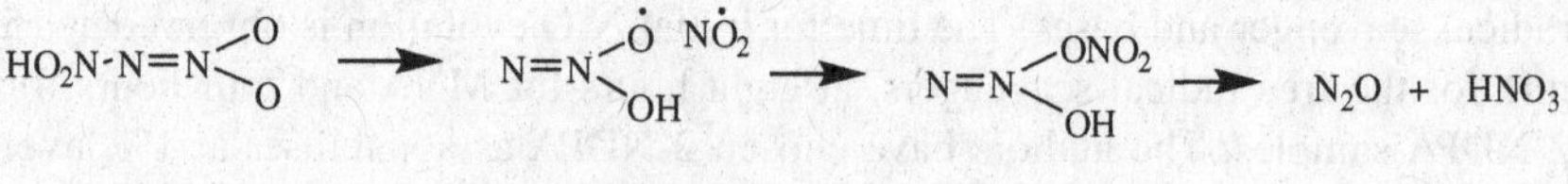

$$HO_2N\text{-}N{=}N\overset{O}{\underset{O}{\diagdown}} \longrightarrow N{=}N\overset{\overset{O}{}\ \dot{N}O_2}{\underset{OH}{\diagdown}} \longrightarrow N{=}N\overset{ONO_2}{\underset{OH}{\diagdown}} \longrightarrow N_2O + HNO_3$$

Scheme 7.4. Decomposition pathway of dinitramidic acid.

$$NH_4N(NO_2)_2 \longrightarrow \{NH_3 + HN(NO_2)_2\} \longrightarrow \{NH_3 + HNO_3 + N_2O\} \longrightarrow NH_4NO_3 + N_2O$$

$$\downarrow$$

$$N_2O + 2H_2O$$

Scheme 7.5. Decomposition pathway of ADN.

of ADN and its additive mixtures were investigated by TGA and DSC under isothermal conditions. They have observed a slow decomposition after the complete melting. The isothermal TGA measurements at 100°C showed a 4% mass loss after 24 hours. Scheme 7.5 shows the decomposition pathway considered for ADN.

From the investigation of various organic and inorganic bases on the stabilisation of ADN, the addition of MgO and NaBO$_2$ showed no changes in the decomposition behaviour, while hexamethylenetetramine, 2-nitrodiphenylamine and akardit II showed a remarkable influence on the decomposition. The TGA experiments show that the onset of decomposition was significantly shifted to higher temperatures. The stability was also confirmed by mass loss tests. The use of organic bases showed higher heats of decomposition, reproducible heat flow spikes and a reduced sublimation of AN after the decomposition. The interaction of ADN with organic bases is attributed to the oxidation of the bases by ADN. The mass loss data obtained from their studies are given in Table 7.1.

Based on their results the effectiveness of stabilisation of ADN with organic bases is in the following order akardit II > 2-nitrodiphenylamine > hexamethylenetetramine.

Boswell *et al.* have studied the thermal stability of ADN[10] and ADN containing various stabilisers. The studies were carried out by microcalorimetry from 80°C to 95°C and by DSC from 115°C to 140°C. Activation parameters were derived for

Table 7.1. Mass Loss Data of Pure ADN and ADN/Stabilisers at 80°C

Sample	Time until a mass loss of 10% / hours
ADN	58
ADN + MgO	55
ADN + NaBO$_2$	97
ADN + hexamethylenetetramine	149
ADN + 2-nitrodiphenylamine	157
ADN + akardit II	248

the decomposition of ADN. The activation energy and rate constants obtained were used to compare the stabilised and unstabilised ADN. Stabilisers which are basic viz. urea, hexamethylenetetramine, cyanoguanidine and tetraethylenepentamine/acrylonitrile (TEPAN) were chosen for the study.

A comparison of the results obtained at isothermal temperature of 120°C for ADN and ADN with urea indicates that the time to show the exotherm is at 58 min and 400 min respectively. The thermal stability of ADN is increased by the addition of urea. The data on the activation energy and rate constant for ADN and ADN with various stabilisers obtained from their study are summarised in Table 7.2.

Their results indicate that the thermal stability of ADN containing urea is high. A comparison of activation energies and rate constants indicated that the urea based composition showed a higher activation energy and a lower rate constant than others. Even though all the samples showed an increase in activation energy, they fail to show a good thermal stability as seen from the rate constant values. A comparison of the rate constant values indicates that hexamine and cyanoguanidine had little or no effect on the stability, while TEPAN shows a decrease in thermal stability.

The liquid phase stabilisation of ADN was studied by Andreev *et al.* by ampoule method following the kinetics of N_2 and N_2O evolution.[11] The kinetics of nitrate ion accumulation was studied spectrophotometrically by measuring the absorbance at 206 nm from which the background absorbance at 284 nm was subtracted. The efficiency of stabilisers was evaluated from the induction point of 120°C for a content of 1 mol% and 1 wt.% additives.

Isotope analysis of ADN labeled with ^{15}N from the NH_4^+ cation showed the absence of ^{15}N in NO and N_2O. This gave the evidence that these fragments were from the anion. The ^{15}N label was found in N_2 in the form of $^{15}N^{14}N$. This gave the conclusion that one N atom originated from NH_4^+ and the other one originated from the anion $N(NO_2)_2^-$. For ADN with a labeled central N atom, the ^{15}N label was found in N_2O which evolved in the form of $^{15}N^{14}NO$. The decomposition mechanism of isotope labeled ADN is shown in Scheme 7.6.

Table 7.2. Activation Energies and Rate Constants for ADN by Isothermal DSC

Sample	E (kJ/mol)	Rate constant at 100°C $(10^{-4}$ min$^{-1})$
Crystalline ADN	138.2	4.2
Prilled ADN/urea	154.9	1.5
Prilled ADN/0.5% hexamine	154.9	3.8
Prilled ADN/0.5% cyanoguanidine	150.7	4.2
Prilled ADN/0.5% TEPAN	146.5	6.8

$$^{15}NH_4N(NO_2)_2 \longrightarrow {}^{15}NN + NO + N_2O$$

$$NH_4{}^{15}N(NO_2)_2 \longrightarrow N_2 + {}^{15}NNO + NO$$

Scheme 7.6. Mechanism of isotope labeled ADN decomposition.

The data on the main decomposition products viz. N_2, N_2O, HNO_3 and AN isotope analysis suggests that the oxidation of ammonium cation is responsible for the formation of N_2, whereas N_2O and NH_4NO_3 are the products of the decomposition of the dinitramide anion. The autocatalysis was inhibited by adding suitable compounds which prevent the oxidation of NH_4^+ cation.

The authors have studied the readily oxidised compounds: amines (both in free states and in composition of complex salts), amides, dinitramide salts, acids, other salts such as KI, KBr, $K_2Cr_2O_7$ and additives such as tetrazole, benzotriazole and 4-amino-1,2,4-triazole. The studied stabilisers decrease the initial decomposition rate by 2–3 fold.

The influence of stabilisers on the decomposition of ADN for some important additives (1 mol %) at 120°C is given in Table 7.3.

Table 7.3. Decomposition Rate of ADN with Selective Stabilisers at 120°C

Stabiliser	Initial decomposition rate/cm^{3}g^{-1}h^{-1}
No additives	6
Diphenylamine	2.2
Hexamethylenetetramine	1.8
m-nitroaniline	2.1
Urea	4.1
Semicarbazide dinitramide	1.8
Dicyandiamide	2.5
Semicarbazide	1.8
Urotropine dinitramide	3.0
Semicarbazide dinitramide	4.5
KI	2.3
$K_2Cr_2O_7$	2.5
$(NH_4)_2SO_4$	2.3
Benzoic acid	3.5
Tetrazole	5.0
Benzotriazole	1.8
4-amino-1,2,4-triazole	2.1

The authors didn't highlight the most suitable stabilisers for ADN. However, from the results given in Table 7.3, it can be inferred that the most efficient stabilisers would be the ones which show the initial decomposition rate below three.

The use of aromatic nitrogen-containing heterocyclic organic compounds for the stabilisation of ADN is disclosed by Ciaramitaro et al.[12] The main highlight of their work is to use purines, pyrimides, pyrazines and triazines substituted with amino, hydroxyl or other activating groups as stabilisers in the weight percent of 0.001% to 5%. The advantage of using these type of stabilisers is that the products of decomposition of ADN reacts with the stabiliser to form a compound which further react at another site on the decomposing ADN molecule. Examples of typical reactions are shown in Scheme 7.7.

Heterocyclic compounds carrying amino or hydroxyl groups gave better stabilisation. The effectiveness of stabilisation depends on number of activating groups attached to the ring. The more the number, better is the stabilisation.

The heterocyclic compounds used as stabilisers tend to form stable charge transfer complexes with the acidic decomposition products of ADN. The authors have given the representative heterocyclic compounds derived from pyridine, pyrimidine, pyrazine, S-triazine, quinoline, quiniazoline, quinoxaline, pyrimidopyridine, pyrimidopyridazine and purine carrying amino and hydroxyl groups in their ring. The authors have studied the decomposition of stabilised ADN compositions by TGA. They revealed that compounds which incorporate amino- or hydroxypyrimidine or aminopyridine functional groups are best suited above all other candidates studied. Weak to moderately strong bases work as stabilisers for ADN. Since the decomposition of ADN takes place at both extremes of pH, moderately basic compounds are identified as suitable candidates.

Mishra and Russell give an account on the thermal stability of ADN[13,14] with stabilisers potassium fluoride (KF), potassium dinitramide (KDN), 6-member

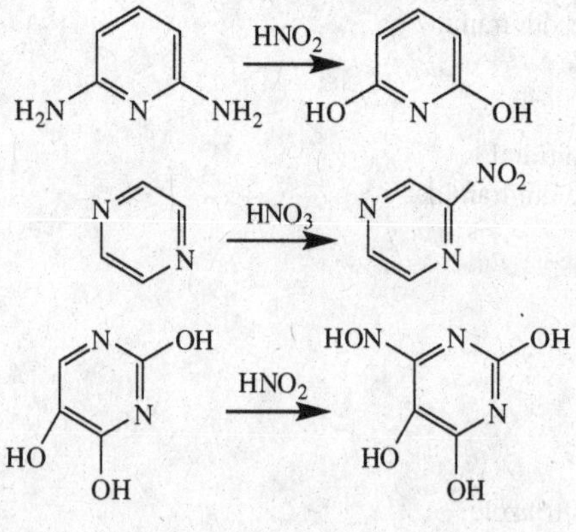

Scheme 7.7. Reaction of the acidic decomposition products on stabilisers.

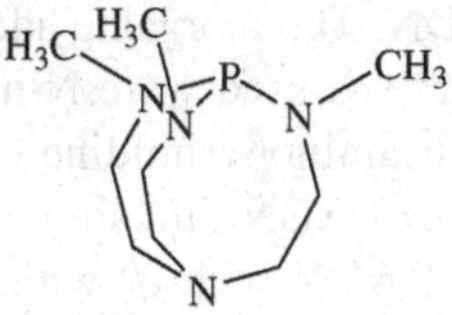

Fig. 7.1. Structure of Verkade's superbase.

ring or polymeric phosphorous compound (PP) $[P(C_6H_5)]_{6(polymer)}$ and Verkade's superbase (VC) [2,8,9-Trimethyl-2,5,8,9-tetraaza-1-phosphabicyclo[3.3.3]unde-cane]. The structure of VC is shown in Fig. 7.1. They have compared the decomposition of ADN at various temperatures at 1%–2% of the mass of stabilisers. A comparative study for the stabilisation of ADN with hexamine was made. The authors have used a heated IR-cell assembly to monitor the gaseous decomposition products N_2O, NH_3 and ammonium nitrate during hourly intervals. The changes in the concentration of N_2O and ammonium nitrate at a constant temperature were employed to evaluate the decomposition and stabilisation effect of ADN.

The stabilisation studies of ADN with KDN, KF, VC and PP at 90°C for the formation of N_2O and AN revealed that VC and KF reduced the amount of AN produced during decomposition, while others enhanced the formation of AN. The authors have identified VC as a suitable stabiliser out of the several stabilisers studied, since it reacts with the formed HNO_3 and consuming it effectively. The effect of light on the decomposition of stabilised ADN with VC was also studied. It was observed that small portions of VC help to stabilise ADN, higher amounts of VC are necessary if the sample is exposed to room light.

A European patent published by Heintz *et al.* describe a method for the production of stabilised and modified spherical ADN particles.[15] The authors have prepared stabilised particles by emulsion crystallisation. The stabiliser is incorporated onto the molten ADN during crystallisation and subsequently cooled to obtain stabilised ADN particles. To avoid the autocatalytic decomposition of ADN, they have used stabilisers such as (1) metal oxides → magnesium and zinc chloride; (2) urea derivatives → urea, N,N-diphenylurea (Akardit I) and N-methyl-N,N-diphenylurea (Akardit II); (3) amine derivatives → hexamethylenetetramine, diphenylamine and nitrodiphenylamine. They have also used purine and its bases such as adenine, guanine, xanthine, uric acid, pyrimidine and its bases, triazine and its derivatives. The authors have successfully demonstrated the migration of the above stabilisers onto the molten matrix of ADN to obtain stabilised spherical particles of ADN.

Zhang *et al.* discuss the thermal stabilisation of ADN with different kinds of organic and inorganic compounds.[16] They have studied the effect of additives on the thermal stabilisation of ADN. The inorganic additives used were MgO, TiO_2 and SiO_2 and the organic additives used were N-methyl-p-nitroaniline, hexamethylenetetramine and 2,4,6-triaminopyrimidine. The authors have explained the decomposition mechanism of ADN in a similar way reported by Lobbecke *et al.*[9] The thermal stability of ADN with or without additives was studied by TGA and DSC techniques. Based on their study, the addition of MgO, SiO_2 and hexamethylenetetramine have no drastic changes on the thermal decomposition of ADN. Whereas the use of N-methyl-p-nitroaniline, 2,4,6-triaminopyrimidine and TiO_2 show a remarkable influence on the decomposition behaviour. The use of stabilisers helps to shift the decomposition temperature to higher range.

Bohn studied the stabilisation of ADN with eight different substances at 2 mol % level.[17] Mass loss as a function of time, isothermal heating and adiabatic self heating were employed to study the stabilisation. The author has not revealed the chemical names of the substances used in the study. Adiabatic self heating studies indicate that ADN is less thermally stable than CL-20, RDX and HMX and the stability was comparable to that of NC. The author has proved that the autocatalytic decomposition was suppressed by the addition of these stabilisers. Various rate equations to represent the stabilisation steps as well as the description of ADN decomposition with mass loss data were given and a rate expression was derived for the autocatalytic decomposition. The author highlights the more important steps in ADN decomposition and compared the activation energy by experimental means with quantum mechanical calculations.

Out of the eight stabilisers studied, seven of them show good results with activation energies in the range of 210 kJ/mol and 280 kJ/mol. All the seven stabilisers help to suppress the protonation of dinitramidic acid. The mass loss as a function of time and temperature at 65°C to 80°C was used to study the decomposition of ADN in the solid state, and the use of accelerating rate calorimetry (ARC) helped to probe the stability of liquid ADN samples at 110°C to 125°C. From the obtained activation energies, the author concludes that the intrinsic decomposition of the anion $N(NO_2)_2^-$ by N–N bond splitting is the rate controlling step with well stabilised ADN samples.

Heintz *et al.* report the preparation of stabilised spherical particles of ADN by incorporating particulate stabilisers inside the ADN prills during the emulsion crystallisation process.[18] The incorporation of particulate stabilisers into ADN prills is achieved by suspending the particulate stabilisers and the crystalline ADN in paraffin oil followed by heating of the oil to 93°C, whereby all the ADN would have been melted. Parts of the stabiliser particles are transported through the phase boundary and incorporated into the ADN droplets.

The authors have not revealed the chemical names of the stabilisers. The stabilised ADN particles were characterised for their sensitivity, particle size, water and ammonium nitrate content. The stability of ADN was measured by heat flow calorimeter, mass loss experiment and vacuum stability tests. The results show

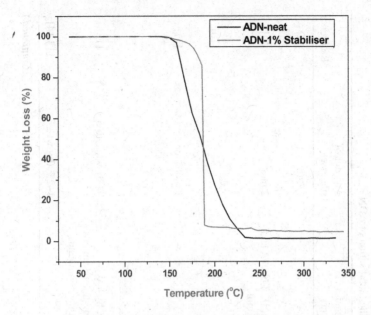

Fig. 7.2. TG traces of neat and stabilised ADN.

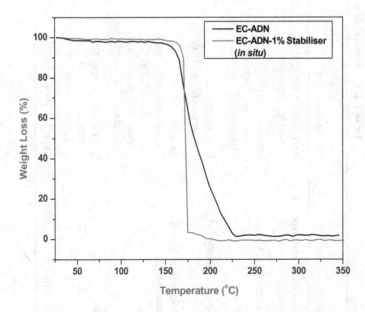

Fig. 7.3. TG traces of emulsion crystallised and *in situ* stabilised ADN.

Table 7.4. Comparative Summary of the Stabilisation of ADN

Techniques used	Stabilisers studied	Suggested stabilisers	Amount of stabiliser	Reference
In situ FT-IR spectroscopy, Isothermal method,	BHT, GAL, DPPH, MNA, 2-NDPA, DPA and NH_3	2-NDPA	1% (by wt.)	[8]
Isothermal TGA and DSC	Magnesia, $NaBO_2$, hexamine, 2-NDPA, akardit II	Akardit II, 2-NDPA and hexamine	2 mol%	[9]
Microcalorimetry and DSC	Urea, hexamine, cyanoguanidine, TEPAN	Urea	—	[10]
Gas evolution by ampoule method	Amines, amides, dinitramide salts, acids, KI, KBr, $(NH_4)_2SO_4$, tetrazole, benzotriazole, 4-amino-1,2,4-triazole etc.,	None	1 mol% or 1 wt.%	[11]
TGA	Heterocyclic compounds carrying amino or hydroxyl groups.	Amino, hydroxy-pyrimidine or amino-pyridine compounds	0.001 wt.% to 5 wt.%	[12]
Heated IR-Cell assembly	KF, KDN, PP and VC	VC	1–2% (by wt.)	[13,14]

(Continued)

Table 7.4. (*Continued*)

Techniques used	Stabilisers studied	Suggested stabilisers	Amount of stabiliser	Reference
None	Metal oxides, urea derivatives, amine derivatives, purine and its bases, pyrimidine and its bases, triazines	None	—	[15]
TGA and DSC	MgO, TiO_2, SiO_2, N-methyl-p-nitroaniline, hexamine, 2,4,6-triaminopyrimidine	N-methyl-p-nitroaniline, 2,4,6-triaminopyrimidine and TiO_2	1%	[16]
Mass loss, isothermal heating and adiabatic self heating	Not mentioned	Not mentioned	2 mol%	[17]
Heat flow calorimeter, mass loss and vacuum stability	Not mentioned	Not mentioned	Up to 1 wt.%	[18]

that pure ADN samples without stabilisers showed autocatalytic early decomposition. The stabiliser embedded ADN showed good thermal stability when subjected to the above tests at 70°C, 80°C, 90°C and 100°C.

Santhosh *et al.* have studied the stabilisation of neat ADN at 1% stabiliser level by TGA.[19] The onset of decomposition for the stabilised mixture was raised by 5°C–10°C at a heating rate of 5°C/min. The TGA plots for the neat and stabilised ADN are shown in Fig. 7.2.

In situ stabilisation of ADN during the emulsion crystallisation (EC) was carried out and it showed that the onset of decomposition was raised by 5°C–8°C at a heating rate of 5°C/min. The TG plots for the emulsion crystallised and the *in situ* stabilised ADN are shown in Fig. 7.3.

The results obtained in our laboratory on the stabilisation of ADN are promising and a detailed study on the same are under way.

7.3. Summary of the Stabilisation of ADN

Table 7.4 gives a comparative summary on the stabilisation of ADN highlighting the stabilisers and their amount, measurement techniques used and suggested stabilisers.

References

1. JC Bottaro, RJ Schmitt, PE Penwell, *et al.,* Method of forming dinitramide salts, *U.S. Patent 5198204,* 1993.
2. JC Bottaro, RJ Schmitt, PE Penwell, *et al.,* Dinitramide salts and method of making same, *U.S. Patent 5254324,* 1993.
3. H Pontius, J Aniol, MA Bohn, Compatibility of ADN with components used in formulations, *35th Int Annu Conf ICT* **169:** 1–19, 2004.
4. Per Sjoberg, Dinitramide News, Eurenco, October 2004.
5. AN Pavlov, GM Nazin, Decomposition mechanism of dinitramid salts. Anomalous decomposition of dinitramid salts and ammonium salt in the solid phase, *29th Int Annu Conf ICT* **25:** 1–14, 1998.
6. GB Manelis, Thermal decomposition of dinitramide ammonium salt, *26th Int Annu Conf ICT* **15:** 1–17, 1995.
7. ZP Pak, AB Andreev, AP Ivanov, *et al.,* Chemical stability of ammonium dinitramide, *Dokl Phys Chem* **375**(1–3)**:** 242–244, 2000.
8. TP Russell, AG Stern, WM Koppes, *et al.,* Thermal decomposition and stabilisation of ammonium dinitramide, *29th JANNAF Combustion Subcommittee Meeting,* CPIA Publication **59**(2)**:** 339–345, 1992.

9. S Lobbecke, H Krause, A Pfeil, Thermal decomposition and stabilisation of ammonium dinitramide (ADN), *28th Int Annu Conf ICT* **112**: 1–8, 1997.

10. RF Boswell, AS Tompa, Thermal stability of ammonium dinitramide (ADN) and ADN containing various stabilisers, *Proc Int Workshop Microcal. of Energ Mat,* Leeds, UK, T-1 to T-7, 1997.

11. AB Andreev, OV Anikin, AP Ivanov, *et al.,* Stabilisation of ammonium dinitramide in the liquid phase, *Russ Chem Bull* **49**(12): 1974–1976, 2000.

12. DA Ciaramitaro, R Reed, ADN stabilisers, *U.S. Patent 6113712,* 2000.

13. IB Mishra, TP Russell, Thermal stability of ammonium dinitramide, *Thermochim Acta* **384**: 47–56, 2002.

14. TP Russell, IB Mishra, Thermal stabilisation of N,N-dinitramide salts, *U.S. Patent 5780769,* 1998.

15. T Heintz, H Krause, U Teipel, Production of stabilised and modified Ammonium dinitramide, *European Patent EP 1331213 A2,* 2003.

16. Zhang Zhizhong, Wang Bezhuo, Ji Yeuping, *et al.,* Thermal stabilisation of ammonium dinitramide (ADN), *Theory and Practice of Energetic Materials* **5**: 259–262, 2003.

17. MA Bohn, Stabilisation of the new oxidiser ammonium dinitramide in solid phase, *8th Int Semin-EUROPYRO* 274–291, 2003.

18. T Heintz, H Pontius, U Teipel, Stabilised spherical particles of ammonium dinitramide (ADN), *35th Int Annu Conf of ICT* **50**: 1–11, 2004.

19. G Santhosh, Ang How Ghee, Unpublished results.

Chapter 8

THERMAL DECOMPOSITION
OF ADN

The thermal decomposition of ADN plays an important role in the combustion characteristics of solid rocket propellants; hence the study of the thermal characteristics is of utmost importance. Many new and innovative technologies are being used for evaluating the decomposition characteristics of materials. The thermal characterisation of ADN is studied by a wide range of thermoanalytical techniques. The phenomenological aspects, kinetics and mechanism of decomposition of ADN are reviewed in this chapter.

8.1. Phenomenological aspects of ADN Decomposition

ADN is thermally stable and has a melting point between 92°C and 95°C. The melting point of ADN is greatly influenced by the presence of impurities. For a 100% ADN, the melting point could be as high as 95°C. The thermal stability of ADN is lower than that of AN and KDN, due to the substantially lower melting point (92°C) and sensitivity towards light. But it is more stable than nitramide (NH_2NO_2) and alkyl dinitramides ($R-N(NO_2)_2$).

DSC and TGA are widely used to get valuable data on the decomposition of ADN under different experimental conditions. DSC can be used to measure a number of characteristic properties of a sample such as melting point, phase transition, decomposition enthalpy, glass transition temperature (T_g) and decomposition temperature. A typical DSC curve of ADN showing the melting and

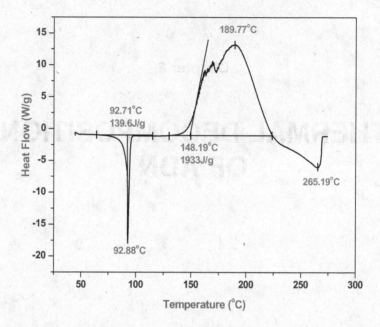

Fig. 8.1. DSC spectrum of ADN at a heating rate of 5°C/min in N_2 atmosphere.

decomposition temperature is shown in Fig. 8.1. The melting point of ADN is 92.8°C, the decomposition exotherm maximum is observed at 190°C with a decomposition enthalpy of 1933 J/g.

Figure 8.1 also showed an endotherm at 265.2°C, which could either be attributed to the formation and vaporisation of water during ammonium nitrate decomposition or the endothermic decomposition/sublimation of *in situ* formed AN. Tompa has studied the effect of different sample pans viz. aluminium, coated aluminium, gold and glass ampoules on the shape of the DSC.[1] The DSC curves of gold and coated aluminium pans have showed a different shape. Much sharper exothermic peaks are observed in sealed glass ampoules. The characteristic decomposition temperature data for ADN and ADN prills by DSC are summarised in Table 8.1.

TGA is performed to determine the change in weight in relation to the change in temperature and is employed to measure the degradation temperatures, weight loss and decomposition steps of a sample under study. The evaluation of the TG trace of ADN shows that it undergoes a single stage decomposition with 100% weight loss without leaving a residue. A typical TG curve of ADN is shown in Fig. 8.2. ADN decomposes in the temperature range of 140°C–240°C.

Table 8.1. Characteristic Decomposition Temperatures of ADN by DSC

Heating rate	Melting point (°C)	Decomp. temperature (°C)	Heat of decomposition (J/g)	Reference
0.5°C/min	92.71	159.3	1970	[2]
2.5°C/min	92.06	182.7	—	[3]
5°C/min	92 ± 1 (crystals)	127.0 ± 2	2100 ± 100	[4]
	90 ± 1 (prills)	133.0 ± 8	2400 ± 400	[4]
5°C/min	92.00	180.0	1960 ± 350	[5]
5°C/min	—	180.0	2508	[6]
5°C/min	94.57	184.0	2023	[1]
5°C/min	92.80 (prills)	182.0	1826	[1]
5°C/min	93.00 (prills)	180.0	1871	[1]
10°C/min	95.37	197.9	2100	[7]
—	93.50	190.1	—	[8]
10°C/min	92.38	185.2	1787	[9]
20°C/min	92.00	189.0	1700	[10]

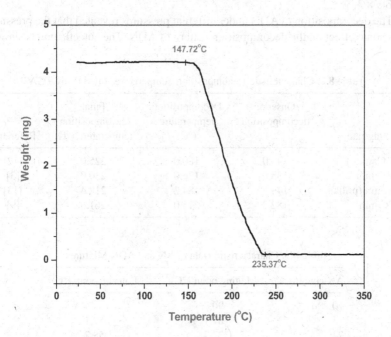

Fig. 8.2. TG trace of ADN at a heating rate of 5°C/min under N_2 atmosphere.

A two stage decomposition of ADN is reported by Lobbecke *et al.* showing a 30% mass loss in the first stage, followed by a 70% mass loss in the second stage (cf. Fig. 4.12). The authors propose the decomposition of ADN to AN and N_2O as the prominent step in the first stage, where the theoretical mass loss for the reaction is 35%. The characteristic decomposition data by TGA obtained for ADN and prills are given in Table 8.2.

Tompa *et al.* point out that when the cation basicity increases, the decomposition temperature increases, while the reaction rate and enthalpy of decomposition decreases.[6] They conclude that the more basic salts were thermally stable and less energetic. The stability of the dinitramide salt may increase as the cation basicity increases.

Ziru *et al.* have studied the thermal decomposition of ADN by pressure DSC and TG techniques.[14] They have derived the kinetic parameters and studied the effect of AN on the decomposition behaviour of ADN. The authors have showed that the presence of AN in ADN strongly affects the melting point of ADN. For a 70/30 mol % ADN/AN mixture, the melting point of the eutectic is as low as 55°C. For AN concentration from 5% to 28%, the melting point of ADN decreases from 87.1% to 76.3°C. The results from their studies are shown in Table 8.3.

The decomposition of ADN under different pressures revealed that the pressure has some effect on the decomposition pattern of ADN. The eutectic points shown

Table 8.2. Characteristic Decomposition Temperatures of ADN by TGA

Heating rate	Onset of decomposition (°C)	Decomposition temperature (°C)	Final decomposition temperature (°C)	Reference
0.5°C/min	140.0	180.0	225.0	[12]
2.5°C/min	135.0	176.0	230.0	[3]
5°C/min (prills)	149.5	184.2	218.4	[13]
10°C/min	182.3	215.0	261.6	[9]

Table 8.3. Characteristic Data of AN and ADN Mixtures

AN content	Melting point (°C)	Eutectic point (°C)
0	90.7	—
5	87.1	59.1
24	77.7	58.7
28	76.3	60.4
29	—	58.8

Table 8.4. Decomposition of ADN with 5% AN Content Under Different Pressures

Pressure (MPa)	Eutectic point (°C)	Peak exothermic temperature (°C)	Enthalpy of decomposition (J/g)
9.3E-03	63.4	176.3 (endo), 195.5	—
0.1	59.1	170.1	1858
1.5	59.4	191.3	2310
3.0	60.2	168.8, 185.3, 205.5	2525
6.0	60.1	171.0, 189.5, 207.4	2653

in Table 8.3 do not vary with the change in pressure; however, the decomposition is greatly affected. Table 8.4 shows the results of ADN/AN decomposition at different pressures.

At low pressure ADN has showed an endothermic decomposition at 176.3°C. With increase of pressure, the endothermic decomposition of AN becomes exothermic and the heat of decomposition increases. The increase of AN content in ADN showed an increase in the decomposition temperature while the heat of decomposition decreases. Russell *et al.* have identified a high pressure monoclinic polymorph (β-ADN) apart from the α-phase.[15,16] Its existence is confirmed by XRD, FTIR, and micro-Raman scattering measurements. The α phase is stable from atmospheric pressure to 2 GPa and it melts above 92°C. Above 2 GPa, the α-ADN transforms into β-ADN, where the transition is temperature independent and the β-phase is stable above 2 GPa in the temperature range −75°C to 140°C. At temperatures above 140°C and between 1 GPa and 10 GPa, ADN undergoes a rearrangement to give ammonium nitrate and N_2O.

The decomposition of ADN in the solid phase was studied by Manelis *et al.*[17] ADN in solid state is 50 times more stable than the same in the molten state. The stability of ADN is greatly influenced by the presence of water and ammonium nitrate. The authors have observed abnormal decomposition behaviour in ADN, whereby the decomposition rate at 60°C is significantly higher than at 80°C, and many times exceeding that of moistened ADN.

ADN has also been studied by isothermal techniques. The isothermal TG traces for ADN prills at 125°C–150°C is shown in Fig. 8.3. The decomposition can be clearly seen from the isothermal TG traces, where the rate of decomposition at 125°C and 135°C are low when compared to 145°C and 150°C. Tompa *et al.* have studied the decomposition of ADN crystals and prills in different environments.[6] The isothermal measurements at 52°C, 62°C and 77°C revealed that a weight loss of 26% was observed at 1400 min–2500 min for samples kept at 52°C and 62°C, while the samples kept at 77°C showed only 4% weight loss during the same time.

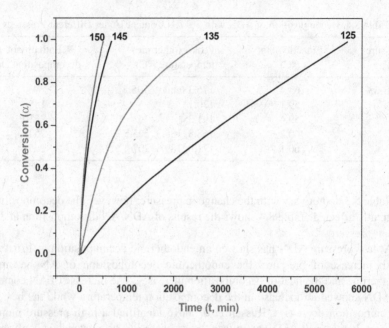

Fig. 8.3. Variation of conversion (α) with time (t) at various temperatures during the isothermal decomposition of ADN prills. The temperature (in °C) is indicated by each curve.

The low temperature thermal stability of ADN was investigated by Mishra *et al.*[18] The thermal decomposition of ADN was studied at 60°C, 70°C, 80°C and 90°C by monitoring the formation of N_2O and AN using heated IR cell method. The decomposition follows a first order reaction. As the temperature increases, the production of N_2O and AN increases, the concentration of AN and N_2O after certain hours tapers off at below 80°C and below 90°C respectively. The amount of N_2O production at early stage is higher at 60°C than at 70°C. The low temperature pyrolysis of ADN was studied by Rossi *et al.*[19] In analogy to NH_4NO_3, the pyrolysis of ADN is dominated by decomposition into NH_3 and $HN(NO_2)_2$ with maximum of decomposition occurring at approximately 155°C.

The decomposition studies of ADN crystals and prills were carried out by Jones *et al.*[4] The authors have used DSC, TG-DTA-FTIR-MS, accelerating rate calorimeter, heat flux calorimeter and isothermal nanocalorimeter for the thermal characterisation. The kinetics and mechanism of thermal decomposition of ADN were studied by Vyazovkin *et al.* by DSC, TG-MS techniques.[5] Brill *et al.* describe the rapid pyrolysis of ADN thin films using T-jump FT-IR spectroscopy.[20] Thin films of ADN (200 μg) were rapidly heated at 2000°C/sec to 260°C. Vyazovkin *et al.* studied the thermal decomposition kinetics of ADN under isothermal and nonisothermal conditions by TGA.[21] Low and high temperature thermal decomposition of ADN are studied by Tompa *et al.*[6] Korobeinichev *et al.* have studied the thermal decomposition of ADN by time-of-flight mass spectrometer at various pressures in the temperature range from 100°C to 300°C.[22] The thermal decomposition of ADN at 1 atm was studied at a constant heating rate of 90°C/sec. Russell *et al.* have studied the thermal decomposition of ADN under isothermal and nonisothermal conditions.[23] *In situ* FT-IR technique was employed for the identification of the decomposition products. Vyazovkin *et al.* have studied the thermal decomposition of ADN using a thin film laser pyrolysis technique.[24] Park *et al.* have studied the low pressure thermal decomposition of ADN by pyrolysis/mass spectrometry.[25] The decomposition of ADN is modeled at 10 Torr using the theoretically computed values of rate constant by *ab-initio* molecular orbital and canonical Variational Rice-Ramsperget-Kassel-Marcus (MO/cVRRKM) calculations.

8.2. Products of Thermal Decomposition

ADN decomposes in the temperature range of 130°C–200°C leaving gaseous decomposition products chiefly of oxides of nitrogen. Products such as N_2O, NO_2, NO, AN, HNO_3, NH_3, $HN(NO_2)_2$ and N_2 are anticipated during the thermal decomposition of ADN and the sequence of evolution and the quantity of these greatly depend on the experimental conditions at which ADN is subjected to. The sequence of evolution of these products was given by many authors in different ways under different experimental conditions. Brill *et al.* studied the decomposition products of ADN using T-Jump FT-IR spectroscopy using very fast heating rates (2000°C/sec) on thin films of ADN in the temperature range from 220°C–300°C.[20] The initial spectrum showed the presence of HNO_3, NH_3 and N_2O in equal amounts along with trace amounts of NO_2, AN and H_2O. During the course of the decomposition, the amount of NH_3 and HNO_3 decreased sharply and more NO_2, N_2O and NO were produced. Russell *et al.* have studied the thermal decomposition of ADN at slow heating rates under isothermal conditions.[23] The authors identified initial decomposition products even at 60°C. N_2O was observed at 60°C, AN sublimation at 80°C and the sublimation of AN along with N_2O is detected at 95°C. At temperatures above 150°C, NO_2 along with N_2O and sublimed AN were detected. At 180°C, the appearance of NH_3 and increase of N_2O, NO_2 and sublimed AN was observed. The products detected for the isothermal decomposition of ADN in the temperature range 60°C–120°C are similar to those performed under nonisothermal conditions. The authors have determined the ADN/AN eutectic by using hot stage optical polarising microscopy.

Rossi *et al.* have carried out the pyrolysis experiments at ambient temperature using a cold cryostat and monitored the decomposition products on a cold KCl window by FT-IR and mass spectrometry.[19] The pyrolysis of ADN at 169°C using the ambient temperature cryostat showed the evolution of NH_3, NO and N_2O together with trace amounts of $HN(NO_2)_2$ [HDN] and the presence of HNO_3 was not detected. At 170°C, the evolution of NH_3 and NO increases and the thermal decomposition of HDN become predominant. The formed HDN began to decompose at 70°C resulting in the rapid evolution of H_2O, NO and N_2O. The ADN pyrolysis studies with a cold cryostat between −160°C to −180°C showed the presence of NH_3 and N_2O in the initial stage together with some NO at a probe temperature of 90°C. At 130°C, NH_3 and N_2O along with H_2O and significant amounts of NO were detected. At 150°C, along with the species identified at 130°C, the presence of HDN is observed. At 170°C, the evolution of HDN reaches maximum, and at 200°C, the evolution of NH_3, H_2O, NO and N_2O is observed at high rates, where the concentration of HDN decreased significantly. At 220°C, none of the species were found. The authors didn't find the presence of possible intermediates NH_4NO_3 or NH_2NO_2. Park *et al.* have studied the decomposition of ADN sublimed at 362K under a pressure of 10 Torr in the temperature range of 373 K–920 K.[25] The authors didn't observe any mass spectral peaks beyond m/z 50. The authors modeled the absolute yields of N_2, N_2O, H_2O and NH_3 during the thermal decomposition of ADN assuming the initial concentration of NH_3 is approximately equal to that of HDN.

The thermal decomposition of ADN by Korobeinichev *et al.* at 1 atm showed the presence of mass spectral peak at 18 in the initial stage and peaks at 17, 28, 30, 44, and 46 emerge in the latter stage.[22] Measurements at 100 Torr at heating rates 90°C/sec showed the presence of N_2, H_2O, N_2O, NH_3, HNO_3 and NO fragments. The thermal decomposition of ADN in high vacuum (10^{-5} Torr to 10^{-7} Torr) showed the mass spectral fragments at 17, 18, 28, 30, 44, 46 and 62. The results indicate that as the pressure decreases, the rate of decomposition of ADN increases and the HNO_3 mole fraction decreases. The formation of NH_3 and $HN(NO_2)_2$ dominate in the initial stage and the decomposition of the latter to N_2O, NO, and H_2O prevails at low pressures. Jones *et al.* using TG-DTA-FTIR-MS technique detected the formation of N_2O, NO_2, H_2O, and CO_2 in the FT-IR spectrum and mass spectral fragments (m/e) at 18, 28, 30, 44 and 46 in the mass spectra of ADN crystals and prills.[4]

The evolved gas analysis performed on the decomposition of ADN by Lobbecke *et al.* showed the presence of AN, N_2O, H_2O, NH_3, NO_2 and HNO_3.[12] The decomposition is initiated by the dissociation of ADN into NH_3 and $HN(NO_2)_2$, while the latter dissociates to produce N_2O and HNO_3. Characteristic IR frequencies at 2234 cm^{-1}–2240 cm^{-1} for N_2O, 1631 cm^{-1}–1635 cm^{-1} for NO_2 and 3176 cm^{-1} –3281 cm^{-1} for AN were observed. Vyazovkin *et al.* detected the formation of N_2O, NO_2, NO, NH_3, HONO and HNO_3 in the decomposition of ADN.[5] The formation of NO_2 was observed at 50°C, followed by the formation of HNO_3. The formation of NO and NH_3 is observed only above 100°C. The initial products detected in the mass spectra are N_2O, NO and NO_2, these are followed at a later

Table 8.5. ADN Decomposition Products at Stoichiometry

Experimental conditions	Stoichiometric equations	Reference
T-Jump FT-IR, Pt filament at 2000°C/sec to 260°C under 1 atm	$ADN \rightarrow$ $0.07NH_3 + 0.22N_2O +$ $0.13NO_2 + 0.33H_2O +$ $0.05NO + 0.13N_2 +$ $0.02HNO_3 + 0.05NH_4NO_3$	[20]
Laser pyrolysis	$ADN \rightarrow$ $0.03NH_3 + 0.29N_2O +$ $0.08NO_2 + 0.31H_2O +$ $0.01NO + 0.13N_2 +$ $0.15NH_4NO_3$	[26]
Thermolysis at 200°C (quenching to 78% decomposition)	$ADN \rightarrow$ $0.03NH_3 + 0.65N_2O +$ $0.68NH_4^+ + 0.22N(NO_2)_2^- +$ $0.46NO_3^-$	[10]

stage by H_2O, HONO, NH_3 and HNO_3. The laser pyrolysis of thin films of ADN with different laser fluences produced N_2O, N_2O_4 (dimer of NO_2), $(NO)_2$ (dimer of NO) fragments as seen by their characteristic absorption bands in FTIR. The stoichiometry of product composition during the thermal decomposition of ADN is summarised in Table 8.5.

8.3.   Kinetics and Mechanism of Thermal Decomposition

The kinetics and mechanism of ADN has been studied extensively by many researchers with reference to temperature, pressure, extent of reaction, thermal cycle, catalysis, presence of additives, isothermal and nonisothermal conditions. Based on experimental conditions and methods, the activation parameters viz. activation energy (E_a) and the pre-exponential factor (ln A) vary.

Ziru *et al.* have calculated the kinetic parameters by DSC for ADN and ADN containing different amounts of AN.[14] The overall activation energy of ADN has increased in the initial stage and decreased in the last stage of decomposition. Oxley *et al.* have calculated the rate constants and activation energy for ADN in the temperature range 160°C–200°C and found E_a of 158.2 kJ/mol for neat ADN and 157.3 kJ/mol and 153.1 kJ/mol for aqueous ADN containing 20% and 1% H_2O respectively.[10] Using the ASTM variable heating rate method, the authors have obtained an activation energy of 113 kJ/mol for ADN. Manelis *et al.* studied the decomposition of ADN in liquid and solid phase.[17] For molten ADN, the decomposition follows a first order reaction above 130°C and at temperatures 100°C–120°C, the reaction has been found to be autocatalytic. The thermal stability of solid ADN is found to be higher than that of liquid ADN. The authors

Table 8.6. Decomposition Mechanism of ADN

Method	Mechanism	Reference
Low temperature decomposition (60–90°C)	$NH_4N(NO_2)_2 \rightarrow (NH_3 + H_3N(NO_2)_2)_{absorbed}$ $(NH_3 + H_3N(NO_2)_2)_{absorbed} \leftrightarrow NH_{3\ (g)} + H_3N(NO_2)_{2\ (g)}$ $(HN(NO_2)_2)_{absorbed} \rightarrow N_2O_{\ (g)} + HNO_{3(g)}$ $HNO_{3(g)} + NH_{3(g)} \rightarrow NH_4NO_{3(s)}$	[18]
DSC and TG-MS	$NH_4N(NO_2)_2 \rightarrow N_2O + NH_4^+ + NO_3^-$ $NH_4N(NO_2)_2 \rightarrow NH_4^+ + NO_2^- + 2\,NO$	[5]
Rapid high rate pyrolysis	$3[ADN \rightarrow NH_3 + HNO_3 + N_2O]$ (low temperature decomposition) $9ADN \rightarrow 6NH_3 + 7N_2O + 10NO_2 + 9H_2O + 3N_2$ (rapid thermolysis) $12ADN \rightarrow 3NH_3 + 10N_2O + 6NO_2 + 15H_2O + 2NO + 6N_2 + HNO_3 +$ $2NH_4NO_3$ (gas phase stoichiometry at the end of exotherm)	[20]
	$NH_4N(NO_2)_2$ (s) $\rightarrow$ $(NH_3 + HN(NO_2)_2)_{absorbed}$ $NH_4N(NO_2)_2$ (s) $\rightarrow$ $(NH_3 + HN(NO_2)_2)_{absorbed} \rightarrow NH_{3(g)} + HN(NO_2)_{2\ (g)}$ $(HN(NO_2)_2)_{absorbed} \rightarrow N_2O_{\ (g)} + HNO_{3(g)}$ $NH_{3(g)} + HNO_{3(g)} \rightarrow NH_4NO_{3(s)}$ $HN(NO_2)_2 \rightarrow NO_2 + HN(NO_2)_2$	[23]
Isothermal TGA and FT-IR	$NH_4N(NO_2)_2 \rightarrow \{NH_3 + HN(NO_2)_2\}$ $\rightarrow \{NH_3 + HNO_3 + N_2O\}$ $\rightarrow NH_4NO_3 + N_2O\ NH_4NO_3$ $\rightarrow N_2O + 2\,H_2O$	[12]
Isothermal TGA and DSC	$NH_4^+\ N(NO_2)_2^- \leftrightarrow NH_3 + HN(NO_2)_2$ $HN(NO_2)_2 \rightarrow N_2O + HNO_3$ $NH_3 + HNO_3 \rightarrow NH_4NO_3 +$ further decomposition products $NH_4^+N(NO_2)_2^- \leftrightarrow NH_4^+ + NO_3^- + N_2O$	[10]
Pyrolysis/mass spectrometry under low pressure	<u>Initiation</u> $HN(NO_2)_2 \rightarrow HNNO_2 + NO_2$ $HNNO_2 + M \rightarrow OH + N_2O + M$ <u>Propagation</u> $OH + NH_3 \rightarrow H_2O + NH_2$ $NH_2 + NO_2 \rightarrow H_2NO + NH$ $NH_2 + NO \rightarrow H + N_2 + OH$	[25]

have obtained an E_a of 140.6 kJ/mol which is equal to the activation energy for molten ADN. The authors have identified an abnormal decay at 60°C, where the decomposition is much higher than at 80°C. The authors have given a mechanism for the decomposition of ADN in the liquid phase and the influence of AN on the stability of ADN at 80°C.

Table 8.6 gives the decomposition mechanism of ADN as postulated by different authors under different experimental conditions.

The thermal decomposition of ADN and ADN prills in different environments was studied by Tompa.[1] Activation energy of 175.3 kJ/mol and 154 kJ/mol was obtained for ADN prills and neat ADN respectively and the frequency factors were 46 min^{-1} and 39.9 min^{-1} respectively as measured by DSC. Santhosh *et al.*

Table 8.7. Kinetic Parameters of Decomposition of ADN

Method	E (kJ/mol)	Ln A (min⁻¹)	Reference
DEA (2°C)	96.0	—	[6]
TGA (5°C)	101.6	24.92	[27]
DSC (ASTM variable heat rate)	112.9	25.92	[10]
TGA (5°C/min)	114.8	32.63	[13]
TGA (2.5°C)	129.5	32.40	[3]
Mass loss (isothermal, 65–80°C)	136.0	33.89	[28]
DSC (isothermal)	138.1	—	[29]
TGA (5.5°C/min)	139.4	35.70	[5]
TGA (isothermal, 125–150°C)	147.4	36.71	[13]
DSC (in N_2)	154.0	39.86	[1]
DSC (0.1MPa) with 5% AN	157.2	40.88	[14]
DSC (ASTM variable heat rate)	158.0	40.48	[30]
Isothermal (160–220°C)	158.2	35.82	[10]
TGA (in He)	168.2	43.00	[1]
DSC (in N_2) Prill	175.3	46.00	[1]
ARC (crystals)	257±9	80±3	[4]
ARC (prills)	310±9	94±3	[4]

have studied the thermal decomposition of ADN and the kinetics was studied by Coats-Redfern, Kissinger and Ozawa methods.[3] The activation energy of ADN is 158.3 kJ/mol, 162.1 kJ/mol and 116 kJ/mol by Ozawa, Kissinger and Coats-Redfern methods respectively. They found that the activation energy for ADN is found to change from 68 kJ/mol to 174 kJ/mol for different conversions as measured by TGA.

Theoretical and experimental studies have been performed on ADN to derive and study the mechanism of thermal decomposition. Table 8.7 summarises the kinetic parameters for the decomposition of ADN.

Isoconversional model-free methods have increasingly become popular during the recent years in evaluating the decomposition kinetics. The activation energy by isoconversional method on the thermogravimetric data for ADN as determined by Vyazovkin et al. varies from 175±20 kJ/mol in the beginning of the reaction and 125 kJ/mol±20 kJ/mol near the completion of the reaction.[5] The authors propose that the initial reaction is the elimination of N_2O from the dinitramide anion and at 200°C, the reaction proceeds via the conversion of ADN to AN. They have also observed that the presence of NO_2 autocatalyses the decomposition of ADN. The dependence of activation energy with conversion for the ADN prills was studied by Santhosh et al.[13] The plots on the dependence of activation energy with conversion computed by Senum-Yang approximation, numerical integration and standard isoconversional methods are shown in Fig. 8.4.

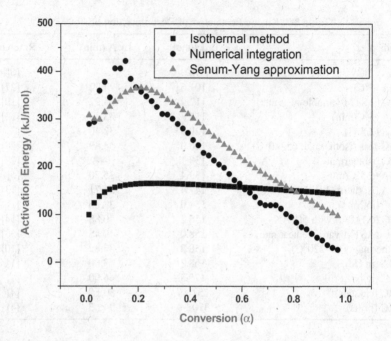

Fig. 8.4. Dependence of activation energy with conversion for ADN prills by different methods.

The dependence can be clearly seen from the figure as the activation energy reached a maximum at 20% conversion and later decreases till the end of conversion. The dependence of activation energy with conversion studied by isothermal method however showed a nearly constant activation energy between the conversions 20% to 100%.

The mass spectral data obtained by Korobeinichev *et al.* at pressures 1 atm and 100 Torr differ from the mass spectral data for the ADN decomposition in vacuum.[22] The observation shows that as the pressure decreases the rate of ADN decomposition increases and HNO_3 mole fraction decreases. The stoichiometric composition of ADN decomposition products as observed by Oxley *et al.*[10] are 0.44 N_2, 0.65 N_2O, 0.68 NH_4^+, 0.22 $N(NO_2)_2^-$, 0.46 NO_3^-. The authors conclude that the mechanism acting on the decomposition of ADN above 160°C is free radical and below 160°C is ionic.

Based on the numerous studies on the thermal decomposition of ADN, one could see that ADN exhibits a two stage decomposition below and above the decomposition temperatures. The condensed phase thermal dissociation of ADN to N_2O and AN or NH_3, N_2O and HNO_3 are responsible for the decomposition at temperatures below 150°C. The formation of NH_3 and HDN or NO_2 and $HNNO_2$ are responsible for the decomposition of ADN at temperatures above 150°C.

References

1. AS Tompa, Thermal analysis of ammonium dinitramide (ADN), *Therm Chim Acta* **357–358:** 177–193, 2000.
2. S Lobbecke, H Krause, A Pfeil, Thermal decomposition and stabilisation of ammonium dinitramide (ADN), *28th Int Annu Conf ICT* **112:** 1–8, 1997.
3. G Santhosh, S Venkatachalam, AU Francis, *et al.,* Thermal decomposition kinetic studies on ammonium dinitramide (ADN) — glycidyl azide polymer (GAP) system, *33rd Int Annu Conf ICT* **64:** 1–14, 2002.
4. DEG Jones, QSM Kwok, M Vachon, *et al.,* Characterisation of ADN and ADN-based propellants, *Prop Expl Pyro* **30**(2): 140–147, 2005.
5. S Vyazovkin, CA Wight, Ammonium dinitramide: Kinetics and mechanism of thermal decomposition, *J Phys Chem A* **101:** 5653–5658, 1997.
6. AS Tompa, RF Boswell, P Skahan, C Gotzmer, Low/High temperature relationships in dinitramide salts by DEA/DSC and study of oxidation of aluminium powders by DSC/TG, *J Therm Anal* **49:** 1161–1170, 1997.
7. JC Bottaro, PE Penwell, RJ Schmitt, 1,1,3,3-tetraoxo-1,2,3-triazapropene anion, a new oxy anion of nitrogen: The dinitramide anion and its salts, *J Am Chem Soc* **119:** 9405–9410, 1997.
8. H Ostmark, U Bemm, A Langlet, *et al.,* The properties of ammonium dinitramide (ADN): Part 1, Basic properties and spectroscopic data, *J Energ Mater* **18:** 123–138, 2000.
9. Zhang Zhizhong, Wang Bezhuo, Ji Yueping, *et al.,* Thermal stabilisation of ammonium dinitramide (ADN), *Theory and Prac of Energ Mater, Vol V, Part A* 259–262, 2003.
10. JC Oxley, JL Smith, W Zheng, *et al.,* Thermal decomposition studies on ammonium dinitramide (ADN) and ^{15}N and ^{2}H isotopomers, *J Phys Chem A* **101:** 5646–5652, 1997.
11. S Lobbecke, M Kaiser, GA Chiganova, Thermal and chemical analysis, in U Teipel (ed.), Energetic materials — particle processing and characterisation, Wiley VCH, 367–401, 2005.
12. S Löbbecke, T Keicher, H Krause, *et al.,* The new energetic material ammonium dinitramide and its thermal decomposition, *Solid State Ionics* **101–103**(2): 945–951, 1997.
13. G Santhosh, Ang How Ghee, Synthesis and thermal characterisation of ADN prills obtained by emulsion crystallisation, in U Teipel, M Herrmann (eds.), Insensitive Energetic Materials — Particles, Crystals, Composites Fraunhofer ICT, 286–299, 2007.
14. Liu Ziru, Luo Yang, Yin Cuimei, *et al.,* Thermal behaviour of a new energetic material ammonium dinitramide, *Int Pyro Semin* 326–333, 1999.

15. TP Russell, GJ Piermarini, S Block, *et al.,* Pressure temperature reaction phase diagram of ammonium dinitramide, *J Phys Chem* **100:** 3248–3251, 1996.
16. TP Russell, PJ Miller, GJ Piermarini, *et al.,* Pressure / Temperature / Reaction phase diagrams for several nitramine compounds, *Mat Res Soc Symp Proc* **296:** 199–213, 1992.
17. GB Manelis, Thermal decomposition of dinitramide ammonium salt, *26th Int Annu Conf ICT* **15:** 1–17, 1995.
18. IB Mishra, TP Russell, Thermal stability of ammonium dinitramide, *Therm Chim Acta* **384:** 47–56, 2002.
19. MJ Rossi, JC Bottaro, DF McMillen, The thermal decomposition of the new energetic material ammonium dinitramide in relation to nitramide and NH_4NO_3, *Int J Chem Kin* **25:** 549–570, 1993.
20. TB Brill, PJ Brush, DG Patil, Thermal decomposition of energetic materials 58. Chemistry of ammonium nitrate and ammonium dinitramide near the burning surface temperature, *Comb & Flame,* **92:** 178–186, 1993.
21. S Vyazovkin, CA Wight, Isothermal and nonisothermal reaction kinetics in solids: In search of ways toward consensus, *J Phys Chem A* **101:** 8279–8284, 1997.
22. O Korobeinichev, A Shmakov, A Paletsky, Thermal decomposition of ammonium dinitramide and ammonium nitrate, *28th Int Annu Conf ICT* **41:** 1–11, 1997.
23. TP Russell, AG Stern, WM Koppes, *et al.,* Thermal decomposition and stabilisation of ammonium dinitramide, *29th JANNAF Comb Subcomm Meeting, Vol. II,* CPIA publication **593:** 339–345, 1992.
24. S Vyazovkin, CA Wight, Thermal decomposition of ammonium dinitramide at moderate and high temperatures, *J Phys Chem* **101:** 7217–7221, 1997.
25. J Park, D Chakraborty, MC Lin, Thermal decomposition of gaseous ammonium dinitramide at low pressure: Kinetic modeling of product formation with *ab initio* MO/cVRRKM calculations, *27th Symp (Int.) on Combustion/The combustion Institute,* 2351–2357, 1998.
26. BL Fetherolf, TA Litzinger, CO_2 laser induced combustion of ammonium dinitramide (ADN), *Comb & Flame* **114:** 515–530, 1998.
27. G Santhosh, S Venkatachalam, K Krishnan, *et al.,* A thermogravimetric study on the thermal decomposition of ammonium dinitramide (ADN) — potassium dinitramide (KDN) mixtures, *34th Int Annu Conf ICT* **16:** 1–8, 2003.
28. MF Bohn, Stabilisation of the new oxidiser ammonium dinitramide in solid phase, *8th Int Semin EUROPYRO 2003,* 274–291, 2003.
29. RF Boswell, AS Tompa, Thermal stability of ammonium dinitramide (ADN) and ADN containing various stabilisers, *Proc of the Workshop on the Microcal of Energ Mater,* Leeds UK, 1–7, 1997.
30. A Langlet, N Wingborg, H Ostmark, ADN: A new high performance oxidiser for solid propellants, in KK Kuo (ed.), *Challenges in Propellants and Combustion, 100 Years after Nobel,* Begell house, UK, 616–626, 1997.

Chapter 9

SENSITIVITY, COMPATIBILITY AND MECHANICAL BEHAVIOUR OF ADN

One of the most important properties for new energetic materials and their formulations is sensitivity. A new energetic material or a new formulation of a propellant has to be thoroughly studied for its sensitivity towards friction, impact and electrostatic discharge and compatibility with other components. The methods and the results obtained on the mechanical sensitivity and compatibility of ADN are reviewed.

9.1. Friction Sensitivity

Langlet *et al.* report the friction sensitivity of ADN measured by using a BAM friction apparatus.[1] A value of > 35 kp was obtained for friction sensitivity. Their results indicate that the sensitivities of ADN are dependant on the crystal shape. The safety tests conducted at China Lake by Chan *et al.* indicate that ADN is slightly more sensitive than HMX and RDX in impact and friction sensitivity.[2] ADN samples with moisture content less than 0.02% showed higher sensitivity. Also samples containing more than 2% hexamine are highly sensitive when compared to neat samples. The results on the friction sensitivities of ADN with hexamine are shown in Table 9.1.

Table 9.1. Friction Sensitivities of ADN

Material	Method	Value	Reference
Neat ADN	ABL friction (50%)		
–0.25% H$_2$O		295 lbs–355 lbs	[2]
–0.5% H$_2$O		295 lbs–355 lbs	[2]
ADN Prills	ABL friction (50%)		
–0.3% Hexamine		295 lbs–355 lbs	[2]
–2% Hexamine		87 lbs	[2]
ADN neat	BAM friction	>35 kp	[1]
ADN neat	—	>350 N	[7]
ADN neat	—	64 N–72 N	[4]
ADN neat	BAM friction (50%)	190 N	[8]
ADN (coated with 1% viton)	BAM friction	48 N	[9]
RDX	BAM friction (50%)	12 kp	[6]
RDX	BAM friction	157 N	[10]

Per Sjoberg *et al*. reported the friction sensitivity of ADN as >350 N.[3] Krause reported the friction sensitivity of ADN as 72 N.[4] Ostmark *et al*. measured the friction sensitivity of ADN[5,6] using a BAM friction apparatus. The friction sensitivity results for ADN (powder) and prilled ADN is >35 kp. The observation shows that the friction sensitivity of ADN is much lower than that of RDX and the measured value is 64 N–72 N.[7] Wingborg *et al*. studied the friction sensitivity of ADN.[8] A summary of the friction sensitivities of ADN is given in Table 9.1.

9.2. Impact Sensitivity

Langlet *et al*. report the drop-weight sensitivity of ADN using a BAM apparatus with a 2 Kg drop weight.[1] The impact sensitivity values were 31 cm and 59 cm for the crystals and prills respectively. Chan *et al*. measured the impact sensitivity of neat and prilled ADN.[2] The drop weight sensitivity of ADN as measured by Sjoberg *et al*. is 31 cm.[3]

The impact sensitivity of ADN is 3 N–5 N as reported by Krause.[4] The BE impact results for a drop height of 102 mm showed that the ADN crystals and prills are somewhat sensitive to impact and are not considered too impact sensitive for transportation in large quantities.[11] The BAM Fallhammer impact energy for ADN crystals and prills showed a value of 4 J in each case. The authors have shown that the impact sensitivity of ADN prills increase when the particle size decreases.

Table 9.2. Impact Sensitivities of ADN

Material	Method	Value	Reference
Neat ADN	BAM — 2.5 kg		
– 0.25% H_2O		11 cm–12 cm	[2]
– 0.5% H_2O		15 cm	[2]
ADN Prills	BAM — 2.5 kg		
– 0.3% Hexamine		11 cm–12 cm	[2]
– 2% Hexamine		11 cm	[2]
ADN neat	BAM apparatus, 2 kg	31 cm	[1,5]
ADN prills	BAM apparatus, 2 kg	59 cm	[1,5]
ADN neat ·	—	31 cm	[7]
ADN neat	—	3–5 N	[4]
ADN neat	BAM apparatus, 2 kg	8 J	[8]
ADN prills	BAM impact	16 J	[10]
ADN (coated with 1% viton)	ERL impact (50%)	6 cm	[9]
RDX	BAM apparatus, 2 kg	38 cm	[6]
RDX	BAM impact	4 J	[10]

Agrawal *et al.* have studied the deformation of ADN by drop-weight method using high-speed photography.[12] Deflagration was observed when ADN was subjected to impact from a drop-height of 1.30 m and it failed to ignite at lower drop heights. The authors have found that ADN is sensitised in presence of 60 μm Pyrex powder, polycarbonate, polystyrene and poly (methyl methacrylate) mixtures. A comparison of the sensitivity results with AP indicates that ADN is slightly more sensitive than AP.

Ostmark *et al.* measured the drop-weight impact sensitivity of ADN using a BAM drop-weight apparatus.[5,6] The drop weight sensitivities for ADN (powder) and ADN prills are 31 cm and 59 cm respectively. The authors have found that the impact sensitivity of ADN is of the same magnitude of RDX and also indicated that the prilled ADN is nearly twice insensitive than the neat ADN. Sjoberg reports the impact sensitivity value of ADN as 3 Nm–5 Nm.[7] The impact sensitivity of neat ADN is classified as 1.1 D according to UN regulations for "Transport of Dangerous Goods".[7] A summary of the impact sensitivities of ADN is given in Table 9.2.

9.3. Electrostatic Sensitivity of ADN

Jones *et al.* have measured the electrostatic sensitivity for ADN crystals and prills.[11] The electrostatic discharge (ESD) results on the ignition energies of ADN

Table 9.3. Electrostatic Sensitivities of ADN

Material	Electrostatic discharge	Reference
Neat ADN		
– 0.25% H_2O	10/10 NF	[2]
– 0.5% H_2O	10/10 NF	[2]
ADN Prills		
– 0.3% Hexamine	10/10 NF	[2]
– 2% Hexamine	10/10 NF	[2]
Neat ADN	2 J (confined)	[13]
	>8 J (unconfined)	[13]
Prilled ADN	9 J (confined)	[13]
	>8 J (unconfined)	[13]
ADN (coated with 1% viton)	ABL ESD — 1.72 J	[9]

samples indicate that the values are greater than 156 mJ and therefore they are not sensitive to ESD. Chan *et al.* have measured the electrostatic sensitivity of neat and prilled ADN.[2] A summary of the ESD values for ADN is given in Table 9.3.

9.4. Detonation Velocity and Detonation Pressure of ADN

The detonation velocity of melt-cast ADN in presence of a stabiliser such as ZnO was measured by Doherty *et al.*[14] The shock wave velocity was read with a Vanguard film reader on a streak camera and found to be 3.42 mm/μs for ADN with 0.05% ZnO. The authors have suggested that the detonation velocity of ADN can be enhanced by the addition of a fuel.

The detonation velocity, pressure and temperature time histories, heat of explosion and mechanical sensitivity of ADN were measured by Gogulya *et al.*[15] The authors have also evaluated the above properties for ADN with aluminium. By measuring the critical pressure of explosion initiation, the sensitivity of ADN to mechanical actions have been measured and the obtained values for crystalline and prilled samples are 0.80 GPa±0.05 GPa and 1.12 GPa±0.04 GPa respectively. The results indicate that the prilled ADN samples are less sensitive than crystalline ADN. The data obtained for the detonation velocity of prilled ADN indicate that ADN belong to group-2 explosives. The density corresponding to the maximum detonation velocity for the prilled ADN samples lies between 1.4 g/cm³ and 1.6 g/cm³. The authors have observed that the increase of charge density

Table 9.4. The Calculated and Experimental Detonation Parameters for ADN

Density of ADN (g/cm³)	Experimental values		Calculated values		Note
	D_{CJ} (mm/μs)	P_{CJ} (GPa)	D_{CJ} (mm/μs)	P_{CJ} (GPa)	
1.568	5.013±0.06	16.0	6.950	16.45	25.1 mm charge
1.658	5.26±0.08	18±2	7.310	18.8	43.9 mm charge
1.820	—	—	7.960	23.6	—

D_{CJ} — Detonation velocity (Chapman-Jouguet)
P_{CJ} — Detonation pressure (Chapman-Jouguet)

results in decrease of detonation pressure and is correlated with the density dependence of detonation velocity. The authors have carried out a detailed study on the detonation velocity, pressure and temperature history for ADN and Al based compositions. Krause reports the calculated values for the detonation velocity and detonation pressure of ADN as 8074 m/s and 23.72 GPa respectively.[4] The calculated detonation parameters for ADN by theoretical and experimental methods are given in Table 9.4.

9.5. Sensitivity of ADN/Binder Mixtures

It is important to study the sensitivity of ADN/binder mixtures as it will provide valuable information of the composition towards mechanical behaviour. Cunliffe *et al.*[10] studied the sensitivity of few ADN-binder systems. The authors have studied the sensitivity of the mixtures at 60% level in the presence of an energetic binder and a plasticiser. The obtained results from their study are shown in Table 9.5.

Compared to other combinations, the ADN/GAP composition is highly sensitive towards friction. The PEG/PPG with BDNPA/F (80%) is less sensitive to friction than that of the PEG/PPG + BDNPA/F (50%).

Table 9.5. Sensitivity of ADN (60%) in Uncured Binders

Binder	Impact energy (J)	Friction sensitivity (N)
Poly NIMMO + BuNENA (33%)	~ 8	~ 60
PEG/PPG + BDNPA/F (50%)	~ 14	~ 70
PEG/PPG + BDNPA/F (80%)	~ 11	~ 90
GAP	~ 14	~ 40

9.6.   Compatibility of ADN

The study of chemical and physical interactions of ADN with other components
used in the formulations is of prime importance and the compatibility should be
studied in detail. Towards more reactive isocyanates, ADN is somewhat incompatible. It also showed some incompatibility with energetic binder PGN. It is
compatible with HTPB and PNIMMO. Results also show that the incompatibility issue doesn't arise for the cured systems. Some of the compatibility data for
ADN with polymers, isocyanates and plasticisers are shown in Table 9.6.

Table 9.6. Compatibility of ADN (1:1) with Various Additives

	Compatibility (J/g)	Reference
Isocyanate		
DDI	18.71	[3], [10]
IPDI	>137	[3], [10]
N-100	>189	[3]
	188.6	[10]
HDI	>385.6	[10]
DNR	>112.1	[10]
$H_{12}MDI$	25.9	[10]
X1004	>125.9	[10]
Binders		
GAP	11.68	[3]
	0.69	[10]
GAP diol	3.1 (8 days)	[16]
PNIMMO	7.34	[3]
	7.35	[10]
HTPB	13.14	[3], [10]
PGN	31.11	[10]
Plasticisers		
Bu-NENA	>320	[3], [10]
BDNPA/F	13.86	[10]
DOS	0.28	[10]
DNEB	0.53	[10]
Oxidizers		
HMX	2.1–4.0 (8 days)	[16]
Curing Catalysts		
DBTDL	4.1	[10]
Fe-octoate	3.5	[10]

(Continued)

Table 9.6. (*Continued*)

	Compatibility (J/g)	Reference
DABCO	44.4	[10]
DMCHA	18.4	[10]
Crosslinking Agents		
TMP	−8.6	[10]
TIPA	18.7	[10]
TEA	25.0	[10]

Pontius *et al.* have studied the compatibility of ADN with few oxidizers and binders.[16] The compatibility was evaluated by mass loss, gas and heat generation techniques. The mass loss data indicated that ADN in the presence of HMX or RDX is quite stable than neat ADN. The gas generation data indicated that ADN with GAP diol, HTPB and DOA showed some reactivity. Nitrocellulose, PGN and PNIMMO are not compatible with ADN where ADN/PGN is extremely reactive with high heat generation. Their studies revealed that ADN is incompatible with compounds containing nitric acid ester groups.

Eldsater *et al.* studied the compatibility of ADN with HMX propellants and some other binders.[17] ADN is chemically compatible with HMX based propellants. It was not compatible with PEG/PPG due to the acidic nature of the polymer. With PTHF, it showed good compatibility. The authors also found that the cured formulations of ADN containing PEG/PPG and BDNPA/F are more stable than the ADN and uncured PEG/PPG. The compatibility results obtained by them are included in Table 9.6.

Cunliffe *et al.* have studied the compatibility of ADN with various additives.[10] The compatibility results are shown in Table 9.6. GAP showed excellent compatibility, PNIMMO and HTPB showed minor incompatibility with ADN, while PGN is totally incompatible. The compatibility of bu-NENA is very poor. BDNPA/F has a minor incompatibility, while DNEB and DOS have excellent compatibility. ADN is more or less incompatible with isocyanates. Only the isocyanates DDI and H_{12}MDI showed good compatibility. DBTDL and Fe-octate are found to show good compatibility and the crosslinking agents under their study proved to be compatible with ADN.

References

1. A Langlet, N Wingborg, H Ostmark, ADN: A New high performance oxidiser for solid propellants, in KK Kuo (ed.), *Challenges in Prop. and Comb. — 100 Years after Nobel*, 616–626, 1997.

2. ML Chan, A Turner, L Merwin, ADN propellant technology, in KK Kuo (ed.), *Challenges in Prop. and Comb. — 100 Years after Nobel*, 627–635, 1997.

3. P Sjoberg, R Wardle, T Highsmith, A cooperative effort to develop manufacturing processes for spherical ADN, *2001 Insens Muni and Energ Mater Tech Symp*, 466–470, 2001.

4. HH Krause, New energetic materials, in U Teipel (ed.), *Energetic Materials — Particle Processing and Characterisation*, 1–25, 2005.

5. H Ostmark, U Bemm, A Langlet, R Sanden *et al.*, The Properties of ammonium dinitramide (ADN): Part 1, basic properties and spectroscopic data, *J Energ Mater,* **18:** 123–138, 2000.

6. S Karlsson, H Ostmark, Sensitivity and performance characterisation of ammonium dinitramide (ADN), *11th Int Det Symp,* 801–806, 1998.

7. P Sjoberg, Chemistry and applications of dinitramides, in PA Politzer, JS Murray (eds.), Energetic Materials, Part I: Decomposition, crystal and molecular properties — *Theor and Comp Chem*, Vol. 12, 389–404, 2003.

8. N Wingborg, Michel van Zelst, Comparative study of the properties of ADN and HNF, FOA-R—00-01423—SE, 2000.

9. GW Nauflett, RE Farncomb, Coating of PEP ingredients using supercritical carbon dioxide, *19th JANNAF Safety & Environ Protec Subcomm Meeting,* **1:** 13–19, 2002.

10. A Cunliffe, C Eldsater, E Marshall, *et al.*, United Kingdom/Sweden collaboration on ADN and polyNIMMO/polyGLYN formulation assessment, FOI-R—0420—SE, 2002.

11. DEG Jones, QSM Kwok, M Vachon, *et al.*, Characterisation of ADN and ADN-based propellants, *Prop Expl Pyro* **30**(2): 140–147, 2005.

12. JP Agrawal, SM Walley, JE Field, High-speed photographic study of the impact response of ammonium dinitramide and glycidyl azide polymer, *J Prop Power,* **13**(4): 463–470, 1997.

13. TK Highsmith, CS Mcleod, R Hendrickson, Thermally stabilised prilled ammonium dinitramide particles and process for making the same, *U.S. Patent 6136115,* 2000.

14. RM Doherty, JW Forbed, GW Lawrence, *et al.*, Detonation velocity of melt-cast ADN and ADN/Nano-diamond cylinders, in MD Furnish, LC Chhabildas, RS Hixson, (eds), *AIP Conf. Proc. Vol. 505 — Shock Compression of Condensed Matter,* 833–836, 1999.

15. MF Gogulya, A Yu Dolgoborodov, MN Makhov, *et al.*, Detonation performance of ADN and its mixtures with Al, *Europyro 2003, 8th Congres International de Pyrotechnie,* 18–27, 2003.

16. H Pontius, J Aniol, MA Bohn, Compatibility of ADN with components used in formulations, *35th Int Annu Conf ICT,* **169:** 1–19, 2004.

17. C Eldsater, N Wingborg, R Sanden, Energetic binders for high performance propellants, FOA-R—00-01610-310—SE, 2000.

Chapter 10

STUDIES ON COMBUSTION
OF ADN

Studies on the combustion chemistry of ADN are very important as it would provide more insight into the understanding of the processes during the combustion and to tailor or improve the performance of solid propellants. To evaluate the combustion characteristics and performance of ADN based propellants, the burning rate and flame structure of ADN has to be studied. The combustion studies of ADN have attracted wide interest in terms of its decomposition, flame structure, mechanism and modeling. This chapter reviews the combustion of neat ADN and its interaction with polymeric binders and additives.

10.1.   Flame Structure of a Solid Propellant

The combustion characteristics of an oxidizer as a monopropellant will have significantly different combustion characteristics in the presence of an energetic binder. A typical solid propellant flame structure proposed by Beckstead is shown in Fig. 10.1.[1]

The condensed phase reactions occur in the thin layer below the surface. In the primary diffusion flame, the propellant burn rate is controlled because of the energy release and proximity to the burning surface. The premixed monopropellant region is pressure sensitive. The final diffusion flame is of secondary importance. The interaction between the premixed and diffusion flame is of prime importance in modeling approach. The decomposition of products of oxidizer can

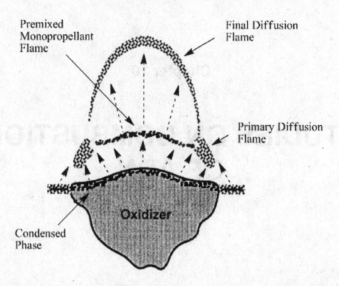

Fig. 10.1. A typical solid propellant flame structure. Reproduced with permission from M. Beckstead.

react in either the diffusion or premixed flame while the decomposition products of the fuel react in diffusion flame.

The chemistry and the study of the heterophase reaction zone near the burning of a propellant are very important as the flame characteristics of a propellant can be derived. During the combustion of a propellant, the very thin condensed phase to gas phase zone has a varying temperature gradient. A mixed phase microstructure of a burning solid propellant is depicted in Fig. 10.2.

It is often difficult to describe the chemistry of the reaction zone in the flame and different approaches were evolved to simulate the conditions of the reaction zone of a propellant by different experimental techniques.

10.2.　Combustion Characteristics of ADN

The burning rate of ADN is higher than that of oxidizers such as AP, AN, HMX, RDX, CL-20 and HNF. The calculated flame temperature for ADN monopropellant is about 2000K. The burn rate of ADN is greatly influenced by particle size, the higher the particle size, the higher is the burning rate. Different analytical tools and methods were developed to study the flame structure and combustion of ADN and these are summarised in Table 10.1.

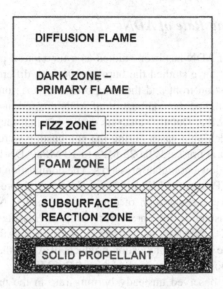

Fig. 10.2. Surface reaction zone of a burning solid propellant. Reproduced with permission from Begell House, Inc. Reference [27].

Table 10.1. Experimental Techniques for the Combustion and Flame Structure Studies of ADN

Sl. No	Technique	Reference
1.	Molecular beam mass spectrometric (MBMS) sampling with time-of-flight mass spectrometer (TOFMS)	[2]
2.	Coherent anti-Stokes Raman scattering (CARS)	[3]
3.	Planar laser induced fluorescence (PLIF)	[4]
4.	Probing mass spectrometry (PMS)	[2]
5.	CO_2 laser induced combustion with microprobe sampling	[5]
6.	T-jump FTIR spectroscopy with high heating rates	[6]
7.	Microthermocouple (MT)	[7]
8.	UV/VIS & IR emission spectroscopy	[8]

The combustion of ADN has been reviewed by Yang *et al.*[9] The authors summarise the combustion wave structures and the burning rate characteristics of ADN. The main features of ADN combustion, the mechanism of combustion in the condensed and gas phase, dependence of burn rate on pressure and issues on the combustion instability are reviewed in the paper.

10.2.1. *Burning Rate of ADN*

The burning rate of ADN has been studied as a function of pressure by Strunin *et al.*[10] The authors have studied the burning rate with different additives, temperature of combustion front and the composition of the combustion products. The authors have studied the combustion of ADN pressed into PMMA tubes up to a pressure of 1 MPa. Detailed analysis on the burning rate, condensable gas products, non-condensable gases during the combustion were performed. Their results indicate that at pressures 0.025 MPa to 10 MPa ADN burns steadily. The burning rate slows down at higher pressures. The combustion temperature at low pressures (0.066 MPa and 0.1 MPa) is lower than at high pressures (2 MPa and 6 MPa). The relative concentration of combustion products NO, N_2O, NO_2 and HNO_3 remains constant at low pressures, while increasing the pressure, the amounts of NH_4NO_3, NH_4NO_2 and $NH_4N_3O_4$ decrease and the amount of H_2O increases. Using the data on combustion products of ADN, the authors have calculated the decomposition reaction of ADN at 0.066 MPa.

Fogelzang *et al.* observed unsteady burning rate in the pressure interval of 2 MPa–8 MPa with burning rates ranging from 26 mm/s to 54 mm/s.[11] The single ADN crystals are incapable of sustained burning in the pressure range of 2 MPa–10 MPa, and showed low burn rates in the pressure intervals of 10 MPa–36 MPa. The authors postulate the cracking of the single crystals as the most likely reason for the above observed phenomena. Pressed samples of ADN showed a burning rate of $0.3 g/cm^2 s$ at 1 atm to $3.4 g/cm^2 s$ at 60 atm.[7] The burning rate data of ADN was generated by Atwood *et al.* in the pressure range from 0.24 MPa to 345 MPa.[12] The burning rates of ADN are high at ambient temperature, while it displayed a scattered behaviour in the pressure range of 2 MPa–10 MPa. The authors attribute this behaviour to the deconsolidation of ADN due to thermal stresses in the sample. The authors have obtained data on the burning of ADN at 298K and 348K over the pressure range of 0.69 MPa–10.3 MPa.

The burn rate data and combustion temperature of ADN at different pressures during the combustion are summarised in Table 10.2.

The burn rate vs. pressure plots for few important oxidizers ADN, HNF, AP and AN as monopropellants are shown in Fig. 10.3.

As it is seen from Fig. 10.3, the burning rate of ADN is high among other oxidizers. ADN showed two slope changes in the burning rate curve indicating a complex combustion mechanism. At the operating pressure of ~ 68 atm, ADN showed a highest burn rate of ~ 3 cm/sec–4 cm/sec.

The characteristic experimental data on the combustion of AP, AN and ADN are summarised in Table 10.3.

As it is seen from Table 10.3, the burning rates of ADN are higher than that of AP and AN.

Table 10.2. Burn Rate and Combustion Temperature Data For ADN

Pressure (atm)	Burn rate (mm/s)	Combustion temperature (K)	Reference
0.99	2.7	560	[10]
1	3.44 ± 0.05	320–370	[2]
3	12 ± 1	550 ± 80	[2]
3	—	1328	[5]
6	19 ± 2	1080 ± 30	[2]
19.7	18.2	700	[2]
40	50–60	1700	[13]

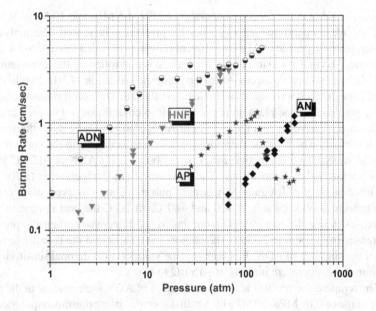

Fig. 10.3. Burning rate vs pressure plots for oxidizers. Reproduced with permission from M. Beckstead.

10.2.2. *Flame Structure of ADN*

The combustion behaviour and flame structure of ADN in the form of pressed strands and single crystals were studied by Fogelzang *et al.*[11] ADN pressed into 4 mm tubes tend to begin self-sustained combustion at 0.2 MPa. The combustion

Table 10.3. Experimental Combustion Data for Selected Oxidizers

	Pressure (atm)	Combustion surface temperature (K)	Burning rate, r (cm/s)	Burning rate exponent, n	Reference
AP	1	770	0.02–0.03	—	[18]
	40	—	0.50	0.77	[18]
	100	—	1.0	0.50	[18]
AN	70	600	0.3–0.6	0.8–0.9	[18]
ADN	1	560	0.27	0.7	[18]
	20	700	1.82	0.7	[18]

process is stable in the pressure intervals 0.2 MPa–1.5 MPa and 10 MPa–36 MPa. The surface temperature increases when the pressure is increased. The analysis of the temperature profile of ADN at different pressures revealed three peculiar zones in flame. The first zone, just above the surface is controlled by the dissociation of AN, the second zone is occupied by the complete oxidation of NH_3, while in the third zone the thermodynamic heat release is attained. The decomposition of ADN in the condensed zone plays an important role in combustion at low pressures.

Zenin *et al.* have studied the physics of ADN combustion by microthermocouple technique, where by the thermocouples were imbedded into the pressed samples of ADN.[7] Two distinctive zones were observed for ADN at −150°C in the pressure interval of 1 atm–60 atm. The maximum flame temperature of 1700°C is obtained at 60 atm. Irregular temperature pulsations are observed when experiments were conducted at T_o = +20 and +80°C. At 20°C the heat release in solid increases, while the heat release from gas to solid decreases when the pressure increases from 1 atm to 20 atm. The authors have measured the thermal sensitivity of burning rate of ADN. It is seen that the values on the thermal sensitivity are high in the temperature interval of −150°C to +20°C.

The temperature profiles and the combustion of ADN were studied in the pressure range of 0.04 MPa–10 MPa by Sinditskii *et al.*[14] using thermocouple method. Temperature measurements carried out on ADN at 0.1 MPa show that it was capable of sustained burning at this pressure and the surface temperature were in the range of 593 K–608 K. Temperature measurements at 0.1 MPa using the 20 μm thermocouple show that the temperature above the burning surface increases slowly, while the molten ADN showed scattered splashes at 650K–700K. The surface temperature increases when the pressure is increased. The temperature measured by using 7 μm thermocouples above the surface of ADN at 2.1 MPa is 845 K–895 K followed by a sharp increase to 1325 K–1375 K at distances of 1 mm–2 mm. At 4.1 MPa, the temperature above the surface rises to 850 K and to 1350 K–1400 K at a distance 0.8 mm–1 mm. The maximum measured temperature did not correspond to the calculated adiabatic flame temperature (2050 K) of ADN. The analysis of the temperature profiles of ADN showed the presence of three distinctive zones in the ADN flame. The authors describe the combustion using a condensed-phase combustion model with decomposition of ADN in the melt as the rate controlling step at low pressures (up to 1 MPa) and the decomposition of HNO_3 as the rate controlling reaction at high pressures (above 10 MPa).

The flame structure of ADN is described by Korobeinichev *et al.*[15] The authors have studied the ADN flame structure at 1 atm–6 atm using MBMS and MT technique. The ADN flame structure involves three zones, at low pressures (1 atm–3 atm) luminous flame was not observed and at 3 atm a cool flame zone adjacent to the burning surface was found.

10.2.3. *Degradation of ADN in the Condensed Phase*

Molecular-beam mass spectrometric technique was applied to study the combustion chemistry of ADN.[13] The method consists of continuously sampling the gaseous species from all the zones of burning by a probe and then transporting it to an ion source of a time-of-flight or quadruple MS. The decomposition of ADN at a heating rate 90°C and 1 atm was studied by MBMS technique and mass peaks at 17 (NH_3^+, OH^+), 18 (H_2O^+), 28 (N_2^+), 30 (NO^+) and 46 (NO_2^+) were observed. Mass peaks at 18 and 44 are first found in the mass-spectrum. The decomposition at 0.132 atm followed the same pattern as that observed at 1 atm. The decomposition study at 7.9×10^{-3} atm showed that more than 90% of ADN got deposited on the cold wall of the tube. Mass peak at 63 was not observed in the spectrum, peaks at 18, 17, 16 and a weak peak at 44 were found. The authors have studied the decomposition of ADN under high vacuum and derived the rate constants. The combustion products showed the following mass spectral fragments: 46 (NO_2, HNO_3), 30 (NO, N_2O, NO_2, HNO_3), 44 (N_2O), 17 (NH_3, H_2O), 28 (N_2, N_2O), 18 (H_2O) and 63 (HNO_3). Based on the detailed studies on the decomposition of ADN at 1 atm, the following species were identified: H_2O, N_2O, NH_3, HNO_3, N_2 and ADN corresponding to the mole fractions 0.33, 0.38, 0.11, 0.09, 0.05 and 0.03. The temperature measurements in the gas phase decomposition of ADN at 40 atm using a thermocouple located close to the strand surface showed the maximum flame temperature as 2000 K.

The condensed phase degradation and combustion of ADN was studied by linear pyrolysis under vacuum.[16] ADN was subjected to linear pyrolysis under vacuum with a regression contact temperature of 370 ± 2 K under 0.3 mbar to 0.5 mbar of pressure. The gases obtained during the combustion NH_3, H_2O, N_2, NO, N_2O, NO_2, HNO_3 and HONO were compared with the results obtained from others.

Fetherolf *et al.* have studied the CO_2 laser induced combustion of ADN in the pressure range of 0.1 atm to 5 atm with laser heat fluxes.[5] The authors have observed three distinctive regions, the first one laser-induced pyrolysis (LIP) addresses the fundamental decomposition characteristics of ADN under low pressure and low heat flux conditions, the second one laser-induced regression (LIR) which is characterised by the absence of luminous flame, gas phase reactions of the decomposition products occur in this region, the third one laser-assisted combustion (LAC) which is characterised by a luminous flame as the ADN regressed under the laser heating. A detailed analysis of the temperature and the profiles of decomposition species for the three distinct regimes were performed by the authors and the results are summarised in the paper. In the LIP conditions, only the formation of N_2O and H_2O and a fair amount of NO_2 and NH_3 along with N_2 and NO were observed after the onset of rapid pyrolysis. The decomposition species observed under LIR were N_2O, H_2O, N_2, NO_2 and NH_3 and some particles of AN. Unstable flame structure was observed in the LAC region, where no AN particles were present and at 3 atm the flame temperature rose to 1350 K.

Brill *et al.* have used T-jump/FTIR spectroscopy to study the pyrolysis of ADN near the burning surface.[6] 20 µg film of ADN was heated at the rate of 2000°C/sec to 260°C, which corresponds to the surface temperature of burning ADN. The analysis of the decomposition products by FT-IR revealed that during the initial stages HNO_3, NH_3 and N_2O appeared in similar amounts. The spectrum also showed the presence of minor amounts of NO_2, AN and H_2O. After the complete decomposition at 2.5 sec, the product ratios were $12N_2O : 6NO_2 : 3AN : 2NH_3 : HNO_3 : NO$. The responsible reactions for the species released during the decomposition of ADN are proposed.

The decomposition of ADN vapor was studied by Shmakov *et al.* in a two-temperature flow reactor.[17] ADN vapours are formed at 80°C–140°C in the first stage of the flow reactor and flows through the second stage of the reactor whereby the vapours can either be condensed or can be decomposed depending on the experimental need. The experimental measurements indicate that ADN forms a molecular complex $[NH_3].[HN(NO_2)_2]$ and then dissociates into NH_3 and $HN(NO_2)_2$. The mass spectral fragments of 17, 18, 30, 44 and 46 were observed in the second stage when the ADN vapour is decomposed in the temperature range of 160°C–900°C. No detection of m/e 63 corresponding to HNO_3 was observed. The authors have determined the rate constants for the ADN vapour dissociation and the Arrhenius parameters. The mass spectrum of ADN obtained was used to interpret the experimental data on the structure of ADN flame at a pressure of 3×10^5 Pa.

Korobeinichev *et al.* have determined the product composition near the burning surface and found that gaseous ADN and dinitramidic acid dominate.[15] Mass spectral peaks 63 (HNO_3^+), 62 ($NH_2NO_2^+$, NO_3^+), 47 (HNO_2^+), 46 (NO_2^+), 45 (HN_2O^+), 44 (N_2O^+), 30 (NO^+), 29 (N_2H^+), 28 (N_2^+), 18 (H_2O^+) and 17 (NH_3^+, OH^+) were found at 1 atm to 6 atm during the combustion. The reaction zone above the

Table 10.4. Mole Fractions of the Combustion Products of ADN During Burning

NH_3	H_2O	NO	N_2O	NO_2	HNO_3	N_2	O_2	Conditions	Reference
0.10	0.26	0.21	0.27	0.11	0.00	0.06	—	[3×10^{-4} to 5×10^{-4} atm]	[16]
0.18	0.17	0.01	0.16	0.38	0.07	0.03	—	1 atm	[2]
0.003	0.44	0.19	0.18	—	—	0.16	0.02	3 atm	[5]
0.08	0.27	0.23	0.23	0.07	0.035	0.08	—	3 atm	[2]
0.08	0.30	0.27	0.13	—	0.08	0.14	—	[3 atm, 3 mm from the burning surface]	[13]
0.012	0.44	0.16	0.18	—	—	0.17	0.02	5 atm	[5]
0.07	0.31	0.23	0.28	—	0.02	0.10	—	[6 atm, 4.4 mm from the burning surface]	[2]

burning surface of ADN is dominated by the ammonia oxidation by nitric acid. The measured combustion temperature is 1400K and the combustion products are H_2, NO, N_2O and N_2. The third zone has a temperature of 2000K.[15] The video recordings of the propellant flame showed several brightly luminous jets of 0.5 mm–1 mm diameter near the burning surface disappearing and reappearing at another site with a lifetime of 0.2 sec. Mass spectral peaks at 17 (NH_3^+), 28 (CO^+, N_2^+), 30 (NO^+), 46 (HNO_3^+, NO_2^+) and 44 (CO_2^+, N_2O^+) were observed.

The mole fractions of ADN combustion products during different experimental conditions are summarised in Table 10.4.

10.3. Combustion Studies of ADN with Additives

Combustion studies involving ADN with various additives were studied by many authors. The studies provide more insight on the combustion characteristics of ADN in presence of various additives used in the formulations.

10.3.1. *Burning Rate of ADN with Additives*

The flame structure characteristics, burning rates and flame temperatures for ADN with different concentrations of HTPB were studied at pressures 0.05 MPa to 0.6 MPa.[19] For ADN/HTPB propellant with 20% and 10% binder content, the burning rate increased from 0.9 mm/s to 1.5 mm/s and further it decreased to 1.2 mm/s for a binder concentration of 3%. Maximum burn rate is achieved for ADN/HTPB propellant with a 10% binder content, while the burn rate of ADN/HTPB propellant with 3% binder content is much lower than that of pure ADN due to the inhibiting effect of small additives. For ADN/HTPB propellant, the burn rate was doubled when the pressure is increased from 0.1 MPa to 0.3 MPa. The burning behaviour of ADN and ADN/Paraffin mixtures were investigated by Weiser *et al.*[8] The burning rate of ADN and ADN/paraffin (90:10) mixtures were measured in the pressure range of 0.1 MPa to 10 MPa. The burning rate of ADN at 2 MPa is close to that of ADN/paraffin mixture and it exceeds above 4 MPa.

The burning behaviour of ADN and its mixtures are given by Weiser *et al.*[20] The combustion properties of ADN with paraffin in the ratio of 90:10 were measured. At 7 MPa, the authors have observed a burning rate of 50 mm/s with a pressure exponent of 0.8. The pressure exponent for the bimodal oxidizer is low (0.72) in comparison with the fine oxidizer (1.0).[21] Korobeinichev *et al.* have studied the burning rate of ADN propellants with CuO, CU_2O, PbO, PbO_2, Pb_3O_4 catalysts and also with oxidizers AP, AN, RDX, HMX and Al.[21] The temperature and the composition of combustion products were also measured. The addition of CuO catalyst increases the rate of reaction near the burning surface and in the condensed phase at 4 MPa.

The combustion of ADN with various organic fuels was studied by Denisyuk *et al.*[22] The burning rate of ADN samples increases in the pressure range of 0.5 MPa–4 MPa and decreases in the 4 MPa to 6 MPa range, then as the pressure is increased further from 6 MPa to 10 MPa, the burning rate slightly increases and upon further increase from 10 MPa to 30 MPa, growth of the burning rate is observed. The burning rate is sensitive and decreases noticeably towards small amounts of additives and moisture. At a pressure of 0.5 MPa, the temperature in the first zone is nearly 950K, and further at a distance of 6 mm from the combustion surface, the temperature is nearly 1300K. The combustion was studied with aromatic hydrocarbons, aliphatic acids, aromatic acids, alcohols, amines and some explosives. The authors conclude that the addition of minor amounts of additives show a decrease in burning rate, at pressures below 1 MPa, and at pressures above 2 MPa, the burning rate of the mixture remains lower than the burning rate of pure ADN.

Combustion studies of ADN propellants were given by Price *et al.*[23] At 300 psi, the burning rate of dry pressed ADN was substantially increased by the addition of ultrafine aluminium powder 'ALEX'. The burn rate of ADN/PBAN sandwiches is higher than that of the neat ADN samples. The authors have studied the pressure dependence of burning rate of sandwiches with different oxidizer laminae and different types of particulate oxidizer in the matrix. The addition of Fe_2O_3 catalyst has no significant effect on the burning rate of ADN/PBAN propellants. The authors conclude that ADN propellant has a high pressure sensitivity than the neat ADN as seen that at 1000 psi, the burning rate is 2.4 times higher than that of neat ADN. The influence of polycaprolactone molecular weight, particle size of ADN, initial temperature and addition of different additives on the burning rate was studied.[21] It was shown that decrease of the polymer molecular weight led to increase of burning rate.

A detailed analysis on the effect of additives on the combustion of ADN is given by Sinditskii *et al.*[24] The authors have studied the effect of small amount of organic compound such as paraffin on the burning rate of ADN. The burning rate vs. pressure plot for ADN doped with paraffin (0.2%) as 4 mm pressed strands with an average density of 1.79 g/cm³ is shown in Fig. 10.4.

The addition of paraffin extends the low pressure limit of sustained combustion of ADN from 0.2 MPa to 0.02 MPa. The burning rate of ADN-paraffin oil (0.2%) coincides with that of ADN in the pressure intervals of 0.2 MPa–2.1 MPa and 15 MPa–36 MPa. In the pressure range of 2 MPa–10 MPa, the ADN-paraffin oil mixture didn't show a scatter in the burn rate. The reason attributed to the behaviour is that paraffin acts as an alternative fuel which is readily oxidisable by the decomposition products of ADN. A comparison of the burn rate of ADN and 90% supersaturated ADN solution revealed that the supersaturated water solution of ADN begins to burn at 1 MPa at a rate less than the burning rate of neat ADN, and as the pressure increases the burn rate of the solution rises rapidly and exceeds the burn rate of ADN at 4 MPa.

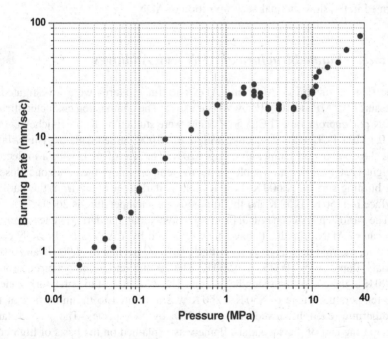

Fig. 10.4. Combustion ADN doped with paraffin (0.2%) as 4 mm pressed strands. Reproduced with permission from V. P. Sinditskii.

Sinditskii *et al.* have studied the combustion behaviour of ADN in presence of organic and inorganic additives, the material of the surrounding shell and sample cross-section.[25] Burning rate measurements were carried out in the pressure range of 0.1 MPa–36 MPa. It was found that ADN burns without any luminous flame at low pressure and a luminous flame is observed at pressures above 1 MPa to 2 MPa. The burning rate data showed a scatter in the pressure range below 10 MPa. By changing the sample diameter or by decreasing the sample density, they could not obtain a sustained burning at atmospheric pressure, while molten ADN samples at temperatures over 100°C were capable of burning at a rate of 6.7 mm/s to 9.1 mm/s. The studies on the effect of combustible shell cover on the burn rate of ADN indicated that it has little influence. The effect of addition of soot, paraffin and silica gel on the burning rate of ADN was studied, the results indicate that these additives decrease the low pressure limit of combustion, while an amount of 1.5% silica gel is able to decrease the pressure limit to a greater extent than the rest. At low pressures, the condensed residue after the combustion of ADN consists of mixtures of AN and ADN with increased ADN content as the pressure decreases. A notable burning rate scatter and a burning rate decrease are observed when the cross-sectional size of the strand is decreased. At atmospheric pressure, the amount of ADN converted to AN (57%) corresponds to the value of AN formed in the slow thermal decomposition of ADN.

10.3.2. *Flame Structure of ADN with Additives*

The flame structure of ADN and ADN/paraffin mixtures were investigated at pressures 0.5 MPa to 10 MPa by Weiser *et al.*[26] The combustion characteristics of ADN/polycaprolactone (PCL) propellants were studied by Korobeinichev *et al.*[21] At 0.1 MPa, PCL with a molecular weight of 10 000 burned without visible flame and the combustion temperature was 670 K. The use of PCL with a molecular weight 1250 has shown a visible flame, the temperature of the dark zone close to the burning surface is 600 K–1150 K and the dark zone away from the burning surface is 1150 K–1450 K and the luminous zone temperature is 2600 K.

The various physicochemical processes involved in the dark-zone temperature plateau of ADN propellants have been reviewed by Yang *et al.*[27] In ADN combustion, the first dark-zone temperature plateau is seen around 900 K and a secondary flame from 1000 K to 1500 K. The second dark-zone occurred around 1500 K at the end of the secondary flame. The existence of the first dark-zone in the temperature range of 850 K–1250 K was attributed to the inhibition of the exothermic reaction between NH_3 and NO_2 by NO species. The second dark-zone at the end of the secondary flame was explained on the basis of high concentrations of NO and N_2O which undergo reaction requiring high activation energies.

10.3.3.   *Thermal Degradation of ADN with Additives*

Temperature profiles of ADN/HTPB propellant with a 3% binder content revealed that the dark zone comprised ADN combustion products.[19] The overall flame structure did not change over a change in pressure. The mass fraction of final combustion products of ADN/HTPB propellant with a binder content 3% are $(0.01)NH_3$, $(0.35)H_2O$, $(0.26)NO$, $(0.18)N_2O$, $(0.13)N_2$, $(0.06)CO_2$.

Weiser *et al.* have recorded the emission spectra of ADN and ADN/paraffin flames in the UV/Vis region and found to have OH, CN and NH radicals.[8] CN and NH radicals are observed in the reaction zone and decrease towards downstream. NH radicals are observed closely above the burning surface. The intensity of OH radicals increases with distance above the burning surface and decreases at the end of the flame. The authors have recorded the IR spectra of the combustion products and found the presence of NO, NO_2 and H_2O in the flame. They have also given a detailed account on the profiles of species and temperature over the burning surface of ADN and ADN/paraffin mixtures. The measured IR spectra showed the presence of H_2O, CO, CO_2 and NO.[26] Addition of 0.2% paraffin to ADN expands the low pressure limit for the sustained combustion from 0.2 MPa to 0.02 MPa.[11] The burning behaviour of ADN-paraffin mixture within the pressure intervals 0.2 MPa–2.1 MPa and 15 MPa–36 MPa coincides with that of pure ADN. Liquified and paraffin doped ADN are capable of sustained burning at 0.1 MPa with a surface temperature of 320°C–335°C.

Using molecular beam mass spectrometry, the flame structure of composite propellants and sandwiches based on ADN and GAP in the pressure range of 0.015 MPa to 0.3 MPa was studied by Kubida *et al.*[28] Two types of ADN and GAP sandwiches were made, where sandwich-I consisted of five 0.8 mm thick ADN laminae and six 0.2 mm laminae of cured GAP, sandwich-II consisted of three 1.6 mm ADN laminae, two 0.4mm GAP laminae between them, and two another 0.2 mm GAP laminae at the ends. Studies were carried out at 0.1 MPa under argon atmosphere using automated mass spectrometric set-up with molecular beam sampling based on a time-of-flight mass spectrometer. Near the burning surface, the following mass spectral peaks were observed: 46 (NO_2 and HNO_3), 18 (H_2O), 30 (NO, HNO_2, NO_2 and HNO_3), 44 (CO_2 and NO_2), 28 (N_2, CO), 16 and 17 (NH_3) and weak peaks at 27 and 29 (HCN and CH_2O). The authors obtain a detailed profile of mass spectral peak intensities for the flame structure of ADN and ADN/paraffin mixtures. The mass spectra of ADN/polycaprolactone has showed spectral fragments corresponding to H_2O, N_2, N_2O, NO, NH_3, HNO_3, H_2, CO, CO_2 and O_2 in the flame.[21]

Tereshenko *et al.* describe a method for the quantitative determination of combustion products at high temperature and pressures.[29] A two-stage probe sampling device is designed which permits the easy sampling of the decomposition products without much change in the composition. The determination of CO and

CO_2 concentrations during the combustion of stoichiometric propellant ADN/polycaprolactone were conducted and their concentrations were determined by mass spectrometric data. The data obtained correspond to the thermodynamic equilibrium concentrations at the measured combustion temperature. The authors suggest that the probe can be used in combustion chambers for the sampling and measurements. Boyarshinov *et al.* used a coherent anti-Stokes Raman scattering (CARS) method to determine the combustion temperature of ADN-polycaprolactone propellant at a pressure of 4 MPa.[3]

10.4.   Modeling the Combustion of ADN

Molecular beam mass-spectrometry and thermocouple measurements were used by Korobeinichev *et al.* to study the ADN flame structure in the pressure range of 0.3 MPa to 0.6 MPa and a mechanism for the chemical reactions in ADN flame was developed based on the experimental and literature data.[30] ADN combustion products NH_3, NO, N_2O, N_2, HNO_3 and H_2O were observed in the mass spectra measured at 0.3 MPa and 0.6 MPa. Based on 172 reactions and 31 species, the first and second zones of ADN flame at 0.3 MPa and 0.6 MPa was modeled. The calculated and experimental temperature profiles are in good agreement each other. The data obtained from their studies can be used for development of combustion model for ADN based propellants.

The combustion mechanisms and flame structure of ADN have been studied by Beckstead.[1] The monopropellant burning rate of ADN has two slope changes in the burning rate curve indicating complex combustion mechanism. The flame temperature of ADN with different weight percent of oxidizer was calculated. The flame temperature of ADN is quite high near its stoichiometric mixture ratio. The author projects the combustion characteristics of propellants based on AP and HMX to the combustion characteristics of advanced solid propellants. The author concludes that the combustion mechanism of ADN appear to follow the mechanisms similar to that of HMX rather than AP propellants.

A multiphase model for the combustion chemistry of ADN is given by Gross *et al.*[31] The authors briefly review the theories appeared in the literature on the combustion of ADN in the condensed and gas phase. Numerical evaluations on the gas phase kinetics, condensed phase mechanisms and evaporation/dissociation of ADN were made. They have obtained results for the first segment of the burn rate curve using a single, global condensed phase reaction model. To develop a model to predict the burning rate in the unstable region (20 atm–100 atm), the authors propose a modification to Sinditskii's theory by a more detailed condensed phase mechanism. They propose the decomposition of AN in the condensed phase to be included in the model. The proposed mechanism suggests that ADN is converted to AN in the first path and then it is converted to NH_3 and dinitramidic acid in the second path, and the majority of AN is assumed to decompose in the condensed phase. The authors provide a more detailed condensed phase reaction mechanism which accounts for all the regions of pressure in the burning rate curve.

The pyrolysis and modeling of the ADN sublimation products under low pressure conditions were given by Ermolin.[32] The pyrolysis of ADN was studied at 10 Torr and the products of the pyrolysis were identified by mass spectrometry. The modeling was done under the assumption that the ADN sublimation products proceeds by formation of ammonia and dinitramidic acid and also by forming ADN vapours. A comparison of the calculated data with experimental data was made and the role of individual stages and the components in the chemical process were also derived.

Modeling and simulation of the combustion of ADN have been carried out by Beckstead *et al.*[33] The solid phase, the condensed phase and the gas phase systems were considered to model the combustion. The insufficient understanding of the condensed phase and the lack of quantitative information about the species leaving the surface of the material have prompted them to develop a model with detailed kinetics. The profiles of temperature and concentration of the decomposition species of ADN at 6 atm were compared with the experimental data and found to be in good agreement.

10.5. Combustion Instability

The dependence of burn rate on pressure has been studied by many authors. The burn rate vs pressure curve of ADN is characterised by three distinct segments, the first one (< 2 MPa) is the increase of burn rate with pressure region, the second one (2 MPa–10 MPa) is the burn rate decrease region and the third one (10 MPa–36 MPa) is again the burn rate growth region. The burn rate in the first segment largely depends on the chemical reactions occurring in the condensed phase and in the third stage, it largely depends on the chemical reactions in the gas phase. Combustion of ADN in the pressure interval 2 MPa–10 MPa (second segment) shows a large scatter of burn rate which clearly indicates the combustion instability. Sinditskii pointed out that ADN crystals are incapable of self sustained burning in the pressure range of 2 MPa–10 MPa.[34] The burning rate vs pressure curve for ADN pressed into 4 mm strands is shown in Fig. 10.5. The unstable combustion of ADN in the pressure range of 2 MPa–10 MPa can be clearly seen by a marked scatter of data.

Sinditskii *et al.* explained the combustion instability with the use of the amount of heat released in the condensed phase. In the pressure range of 2 MPa–10 MPa, the heat released in the condensed phase and the heat flux from the gas phase is insufficient to warm up the surface temperature and to warm up the condensed

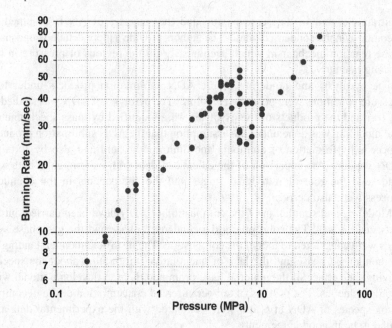

Fig. 10.5. Combustion of pure ADN as 4 mm pressed strands. Reproduced with permission from V. P. Sinditskii.

phase respectively. For want of heat, the combustion in the second region becomes unstable as seen from the burn rate data scatter. Fogelzang *et al.*[11] inferred that cracking of the ADN crystals during ignition and combustion is the most likely reason for the burning behaviour in the second region of the pressure vs. burn rate curve. The authors also postulate that if the surface temperature of ADN follows the AN dissociation, combustion instability should happen. At the pressure range of 2 MPa–10 MPa, the massive formation of AN on the burning surface of ADN can be correlated to the combustion instability. The higher amount of AN in the propellant burning surface is not advantageous, and as the pressure increases, the decomposition of ADN is accelerated and a higher burning rate is achieved.

References

1. (a) MW Beckstead, An overview of combustion mechanisms and flame structures for advanced solid propellants, *36th AIAA/AMSE/SAE/ASEE Joint Prop Conf and Exhibit*, AIAA-2000-3325, (b) MW Beckstead, Overview of combustion mechanisms and flame structures for advanced solid propellants, in V Yang, TB Brill, W-Z Ren (eds.),

Solid Propellant Chemistry: Combustion, and Motor Interior Ballistics, *Prog Astro Aero* **185**: 267–285, 2000.

2. OP Korobeinichev, LV Kuibida, AA Paletsky, *et al.*, Study of flame structure, kinetics and mechanism of the thermal decomposition of solid propellants by probing mass spectrometry, in KK Kuo (ed.), *Challenges in Propellants and Combustion: 100 Years after Nobel*, Begell House, 38–47, 1997.

3. BF Boyarshinov, SYu Fedorov, Measurement of the combustion temperature of a solid propellant by the CARS method, *J Appl Mech Tech Physics*, **43**(6): 925–929, 2002.

4. TP Parr, DM Hanson-Parr, Solid propellant flame structure, *Mat Res Soc Symp Proc* **418**: 207–219, 1996.

5. BL Fetherolf, TA Litzinger, CO_2 laser induced combustion of ammonium dinitramide (ADN), *Combust Flame* **114**: 515–530, 1998.

6. TB Brill, PJ Brush, DG Patil, Thermal decomposition of energetic materials 58. Chemistry of ammonium nitrate and ammonium dinitramide near the burning surface temperature, *Combust Flame* **92**: 178–186, 1993.

7. AA Zenin, VM Puchkov, SV Finjakov, Physics of ADN combustion, *37th AIAA Aerospace Sciences Meeting and Exhibit*, AIAA-99-0595, 1999.

8. V Weiser, N Eisenreich, A Baier, *et al.*, Burning behaviour of ADN formulations, *Prop Expl Pyro* **24**: 163–167, 1999.

9. R Yang, P Thakre, V Yang, Thermal decomposition and combustion of ammonium dinitramide, *Comb Expl Shock Waves* **41**(6): 657–679, 2005.

10. VA Strunin, AP D'yakov, GB Manelis, Combustion of Ammonium Dinitramide, *Combust Flame* **117**: 429–434, 1999.

11. AE Fogelzang, VP Sinditskii, VY Egorshev, *et al.*, Combustion behaviour and flame structure of ammonium dinitramide, *28th Int. Annu Conf ICT* **99**: 1–14, 1997.

12. AI Atwood, TL Boggs, PO Curran, *et al.*, Burning rate of solid propellant ingredients, Part 1: Pressure and initial temperature effects, *J Prop Power* **15**(6): 740–747, 1999.

13. (a) OP Korobeinichev, LV Kuibida, AA Paletsky, *et al.*, Molecular-beam mass-spectrometry to ammonium dinitramide combustion chemistry studies, *J Prop Power* **14**(6): 991–1000, 1998. (b) OP Korobeinichev, LV Kuibida, AA Paletsky, *et al.*, Development and application of molecular beam mass-spectrometry to the study of ADN combustion chemistry, *36th Aerospace Sciences Meeting & Exhibit*, AIAA 98–0445, 1998.

14. VP Sinditskii, VY Egorshev, AI Levshenkov, *et al.*, Combustion of ammonium dinitramide, Part 2: Combustion mechanism, *J Prop Power* **22**(4): 777–785, 2006.

15. OP Korobeinichev, Flame structure of solid propellants, in V Yang, TB Brill, W-Z Ren (eds.), *Solid Propellant Chemistry, Combustion and Motor Interior Ballistics*, *Prog Astro Aero* **185**: 335–354, 2000.

16. J Hommel, Jean-Francois Trubert, Study of the condensed phase degradation and combustion of two new energetic charges for low polluting and smokeless propellants: HNIW and ADN, *33rd Int Annu Conf ICT* **10**: 1–17, 2002.

17. AG Shmakov, OP Korobeinichev, TA Bol'shova, Thermal decomposition of ammonium dinitramide vapor in a two-temperature flow reactor, *Comb Expl Shock Waves* **38**(3): 284–294, 2002.

18. GB Manelis, GM Nazin, YuI Rubtsov, *et al.*, Combustion of pure substances: Reaction in the condensed phase, in *Thermal Decomposition and Combustion of Explosives and Propellants*, Taylor & Francis, Chapter 18, 271, 2003.

19. OP Korobeinichev, AA Paletsky, Flame structure of ADN/HTPB propellants, *Combust Flame* **127**: 2059–2065, 2001.

20. V Weiser, N Eisenreich, A Bayer, *et al.*, Burning behaviour of ADN-mixtures, *28th Int Annu Conf ICT* **8**: 8–14, 1997.

21. (a) OP Korobeinichev, AA Paletsky, AG Tereschenko, *et al.*, Study of combustion characteristics of ammonium dinitramide/polycaprolactone propellants, *J Prop Power* **19**(2): 203–212, 2003. (b) OP Korobeinichev, AA Paletsky, AG Tereshenko, *et al.*, Study of combustion characteristics of the ADN-based propellants, *32nd Int Annu Conf ICT* **123**: 1–14, 2001. (c) OP Korobeinichev, EV Volkov, AA Paletsky, *et al.*, Flame structure and combustion chemistry of ammonium dinitramide/polycaprolactone propellant, *33rd Int Annu Conf ICT* **104**: 1–14, 2002.

22. AP Denisyuk, TM Kuleshova, Yu G Shepelev, Combustion of ammonium dinitramide with organic fuels, *Dokl Phys Chem* **368**(1–3): 271–273, 1999.

23. (a) EW Price, SR Chakravarthy, JM Freeman, *et al.*, Combustion of propellants with ammonium dinitramide, *34th AIAA/ASME/SAE/ASEE Joint Propulsion Conference and Exhibit*, AIAA 98–3387, 1998. (b) SR Chakravarthy, JM Freeman, EW Price, *et al.*, Combustion of propellants with ammonium dinitramide, *Prop Expl Pyro* **29**: 220–230, 2004.

24. VP Sinditskii, AE Fogelzang, VY Egorshev, *et al.*, Combustion peculiarities of ADN and ADN-based mixtures, in KK Kuo, LT Deluca (eds.), *Combustion of Energetic Materials*, Begell House, 502–512, 2002.

25. VP Sinditskii, VY Egorshev, AI Levshenkov, *et al.*, Combustion of ammonium dinitramide, Part 1: Burning behaviour, *J Prop Power* **22**(4): 769–776, 2006.

26. V Weiser, N Eisenreich, S Kelzenberg, *et al.*, Experimental and theoretical investigation of ADN model propellant flames, *37th AIAA/ASME/SAE/ASEE Joint Propulsion Conference and Exhibit*, AIAA 2001–3857, 2001.

27. R Yang, P Thakre, Y-C Liau, *et al.*, Formation of dark zone and its temperature plateau in solid-propellant flames: A review, *Combust Flame* **145**: 38–58, 2006.

28. LV Kubida, OP Korobeinichev, AG Shmakov, *et al.*, Mass spectrometric study of combustion of GAP- and ADN-based propellants, *Combust Flame* **126**: 1655–1661, 2001.

29. AG Tereshenko, OP Korobeinichev, PA Skovorodko, *et al.*, Probe method for sampling solid-propellant combustion products at temperatures and pressures typical of a rocket combustion chamber, *Comb Expl Shock Waves* **38**(1): 81–91, 2002.

30. OP Korobeinichev, TA Bolshova, AA Paletsky, Modeling the chemical reactions of ammonium dinitramide (ADN) in a flame, *Combust Flame* **126**: 1516–1523, 2001.

31. ML Gross, MW Beckstead, KV Puduppakkam, *et al.,* Multi-phase combustion modeling of ammonium dinitramide using detailed chemical kinetics, *42nd AIAA/ASME/SAE/ASEE Joint Propulsion Conference and Exhibit,* AIAA 2006–4747, 2006.

32. NE Ermolin, Modeling of pyrolysis of ammonium dinitramide sublimation products under low-pressure conditions, *Comb Expl Shock Waves* **40**(1): 92–109, 2004.

33. MW Beckstead, KV Puduppakkam, V Yang, Modeling and simulation of combustion of solid propellant ingredients using detailed chemical kinetics, *40th AIAA/ASME/SAE/ASEE Joint Propulsion Conference and Exhibit,* AIAA 2004–4036, 2004.

34. VP Sinditskii, Reason for heat instability of combustion of energetic materials with condensed phase leading reaction, *Cen Eur J Energ Mat* **2**(1): 3–20, 2005.

Chapter 11

APPLICATIONS OF ADN AND DINITRAMIDE SALTS

The combustion of ADN unlike AP doesn't release hydrogen chloride into the atmosphere during the launch applications. This is of practical importance from environmental perspective. Composite propellants based on ADN will outperform in terms of performance than propellants based on AP. With lower solid loading, theoretical calculations indicate that propellants based on ADN/energetic binder will have a higher specific impulse. Taking into consideration the advantages of ADN, the research is focusing on the application and use of ADN in various programs with special emphasis on its use as a solid rocket propellant oxidizer. The relatively large number of other dinitramide salts apart from ADN also finds a variety of applications. This chapter reviews the very many important applications of ADN and other dinitramide salts.

11.1. Low Signature Rocket Propulsion

It was learnt that in the former Soviet Union, ADN has been used in strategic missiles in as early as 1971, but details on the specific characteristics and formulations are still not known to the scientific community. ADN is recognised as an energy-rich oxidizer for use in minimum signature solid propellants. Theoretical calculations show that propellants based on ADN and energetic binders will have 8%–10% increase in lift capacity depending on the application. The use of ADN in formulations will improve the overall energy and performance of a solid rocket motor.

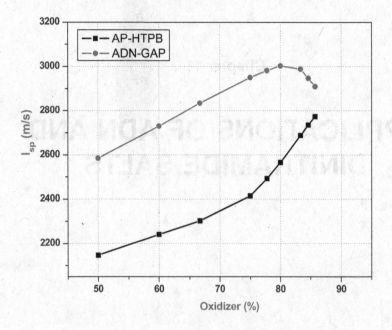

Fig. 11.1. Theoretical performance of ADN based pseudo-propellant.

Figure 11.1 shows the theoretical performance of pseudo-propellants based on ADN-GAP in comparison to AP-HTPB. It can be seen that the specific impulse is much higher for ADN based pseudo-propellant.

Fawls *et al.* disclosed a propellant composition comprising the use of nano size boron with ADN.[1] A gelled propellant was made comprising a liquid fuel, ADN and about 2% to 46% boron particles. Composite propellants based on ADN with energetic binders were disclosed.[2] ADN in combination with PGN and a metal fuel were made and tested. In comparison with AP based propellants, the new formulations offered 3%–9% higher specific impulse. Composite propellants were made with ADN and energetic binders based on oxetane or oxirane polymers in the presence of a metallic fuel. These compositions have high performance, high burn rate and produce minimum smoke without releasing any HCl.

A minimum smoke propellant based on ADN and GAP has been tested.[3] A 75% ADN and 25% GAP mixture was used to test the propellant in a ballistic test motor. The propellant showed a higher specific impulse, low glass temperature (−36°C) and good thermal stability. The burning of ADN/GAP propellant during a test firing is shown in Fig. 11.2. It can be seen that the burning of ADN-GAP propellant is absolutely smoke free.

Fig. 11.2. ADN/GAP propellant during test firing. Courtesy of Swedish Defence Research Agency, FOI.

11.2. High Performance Underwater Explosive

ADN has a detonation velocity of 7000 m/s and can be regarded as a high explosive. Therefore it can be used as an oxidizer in underwater explosives. High explosives such as TNT and RDX don't contribute with much oxygen for burning the fuel, while ADN can supply more oxygen for the complete burning of the fuel. Since ADN has a low melting point, it is possible to prepare melt cast charges for underwater explosives. A melt casting method was developed utilising ADN and ADN/Al with 1% magnesium oxide as a stabiliser.[4] The density of ADN/MgO (99/1) and ADN/Al/MgO (64/35/1) was 92%–97% and 95%–99% of the theoretical mean density (TMD). These explosive charges can find use in underwater explosives.

11.3. Liquid Monopropellants

For commercial and military use, hydrazine has been the standard liquid monopropellant since 1960s. Increased environmental awareness and safety issues in handling hydrazine made the propellant community to find other liquid

Table 11.1. Performance of Liquid Monopropellants

System	Density (g/cm³)	T (°C)	I_{sp} (sec)	LD50 rat (mg/kg)
ADN/glycerol/water	—	—	246.9	1360 (pure ADN)
ADN/methanol/26% H_2O	1.3	1730	253.0	—
HAN/glycine/water	—	—	204.2	325 (pure HAN)
Hydrazine	1.0	1120	230.0	59

monopropellant alternatives. Hydrazine is not only toxic but also carcinogenic. Over the years nontoxic liquid monopropellants based on ADN have been experimented by scientists all over the world. In terms of specific impulse, ADN based monopropellant outperforms hydrazine. A comparison is shown in Table 11.1.

As it is seen from Table 11.1, the specific impulse of ADN based monopropellants is higher than that of HAN and hydrazine based. In terms of toxicity (LD50), ADN is less toxic compared to that of HAN and hydrazine.

ADN based monopropellant is being tested for use in spacecraft propulsion by Swedish Space Corporation (SSC). Monopropellant engines use the catalytic decomposition of energetic liquids into lighter gaseous decomposition products. In the late 1970s, hydroxyl ammonium nitrate has been regarded as a replacement for hydrazine based monopropellants. But the specific impulse of monopropellants based on HAN was not higher than that of hydrazine.

Van den berg *et al.* disclosed a monopropellant based on ADN and water.[5] The use of 50% ADN/water mixture showed higher performance than hydrogen peroxide and hydrazine based compositions. The authors have also used lower alcohols such as methanol and ethanol with ADN. A novel storable liquid monopropellant based on ADN has been disclosed by Anflo *et al.*[6] A ternary formulation has been achieved using ADN along with a fuel and water. The efficiency of the monopropellant combination has been verified experimentally on few 1 N-class rocket engine assemblies. A specific impulse of 240 sec has been achieved with the experimental rocket engine. SSC and Volvo Aero are working together to harden the rocket nozzle with rhenium and other additives to withstand the high combustion temperature of ADN propellants.[7]

At the Swedish Defence Research Agency, FOI, a nontoxic green monopropellant based on ADN has been developed and tested.[8] Two compositions based on 65.7% ADN//13.6% water//20.7% fuel and 64.6% ADN//23.9% water//11.5% fuel were developed and tested. When compared to hydrazine, the above compositions offered 10% higher specific impulse and 60% higher density impulse than hydrazine. In order to put them in use, these ADN based monopropellants are thoroughly investigated for the specific heat capacity, density, and electrical conductivity. An electrical ignition arrangement has been made to ignite the monopropellant with minimal energy; very fast ignition rate of less than 2 ms was obtained.

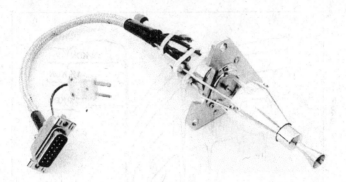

Fig. 11.3. High performance rocket engine assembly/thruster for ADN based monopropellants. Reproduced with permission from T-A. Grönland [Reference 6].

A typical rocket thruster for ADN based monopropellant application developed at the Swedish Space Corporation is shown in Fig. 11.3.

11.4. Explosive Compositions

The high performance of ADN is also of interest in tactical missiles for example to increase the performance of a charge in the warhead. The advantage of using ADN would be to reduce the smoke trail left during the launch, which would significantly increase the safety of the missile launch platform. The fairly low melting point of 92°C is advantageous in producing ADN/Al composites using melt cast techniques similar to TNT based compositions.

A binary high explosive composite based on ADN with hexanitrohexaza-isowurtzitane (HNIW), trinitroazetidine (TNAZ) and trinitrobenzene (TNB) has been disclosed by Langlet *et al.*[9] A melt cast technique was used to prepare ADN-HNIW, ADN-TNAZ and ADN-TNB mixtures. Different proportions of ADN with these explosives were made and analysed. A melt cast charge of ADN with RDX, HMX and HNIW was disclosed.[10] It was found that a mixture of ADN/HMX (30/70) has the same performance of HMX (100). Similarly, the melt cast charges made with RDX and HNIW also show high performance than explosive charges made with TNT.

11.5. Phase-stabiliser in Ammonium Nitrate

It has been shown that metal salts of dinitramides inhibit the low temperature phase transition of AN occurring at about 30°C. The low temperature phase transition in AN would greatly affect its use as an oxidizer in propellants and other

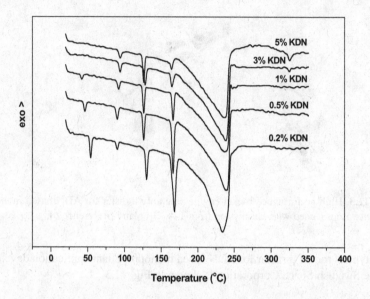

Fig. 11.4. DSC traces of PSAN with different amounts of KDN.

explosive compositions. Apart from the normal phase stabilisers such as KF and KNO_3, the use of metal salts of dinitramides will add to the total energy of the formulations. Santhosh *et al.* have studied the use of KDN as a phase stabiliser for AN.[11] The use of up to 3% KDN (by weight) has been shown to suppress the unwanted IV-III phase transition. The DSC overlay of phase stabilised ammonium nitrate (PSAN) with different amounts of KDN are shown in Fig. 11.4.

Kim *et al.* disclosed a method for producing PSAN using KDN.[12] To a saturated mixture of 65.7% AN, 5.7% KDN and 28.6% water was added MeCN and the precipitated PSAN was removed and used. Phase stabilisation of ammonium nitrate using various metal dinitramide salts are disclosed by Highsmith *et al.*[13] The PSAN was obtained by mixing AN and a metallic dinitramide salt such as potassium dinitramide, cesium dinitramide, lithium dinitramide or zinc dinitramide in methanol and warming the solution to 60°C and evaporating the resultant solution under vacuum. The stabiliser concentration in AN ranged from 5 wt% to 15 wt %. The authors have also used mixtures of metallic dinitramide salts for the stabilisation.

11.6. Reagent in Chemical Synthesis

Few of the dinitramide salts find use as reagents in chemical synthesis. The potassium salt of dinitramide $KN(NO_2)_2$, is an excellent fluorine-oxygen reagent

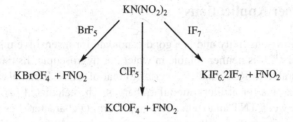

Fig. 11.5. Reaction of potassium dinitramide with fluorine compounds.

capable of reacting with BrF_5, ClF_5 and IF_7.[14] The reaction products are schematically shown in Fig. 11.5.

Mercury salt of dinitramide has been found to be a useful reagent in the chemistry of organomercury compounds.[15] The mercury-dinitramide bond in $Hg(N_3O_4)_2$ is covalent where the Hg atom is bound to oxygen atoms of the nitro group. The compound is readily soluble in water. Several types of reactions such as alkylation, mercuration, complexation and addition with olefins were studied.

ADN finds use in metathetical reactions to synthesise anhydrous cesium salts.[16] The reaction of ADN with cesium methoxide in methanol afforded the cesium salt along with volatile product $NH_4^+CH_3O^-$ as shown in Scheme 11.1.

$$NH_4^+N(NO_2)^- + CsOCH_3 \xrightarrow{\quad CH_3OH \quad} CsN(NO_2)_2 + [NH_4^+ CH_3O^-]$$

Scheme 11.1. Metathetical reaction of ADN with cesium methoxide.

The reaction yields very pure anhydrous salts in quantitative yields.

11.7. Biological Applications

Dinitramide salts were used to increase the solubility of biologically active agents in lipophilic media.[17] The solubility of hydrophobic pharmacologically active agents can be increased by the use of N,N-dinitramide salts. This can be of use in pharmaceuticals, drug delivery, medical imaging and agrochemicals. The admixture of an ionisable compound with a dinitramide salt in a lipophilic medium forms an ionisable compound carrying the anionic counter ion from the dinitramide salt which becomes biologically active. Different dinitramide salts can be employed for increasing the solubility of the biologically active agents and ADN finds use in such applications.

11.8. Other Applications

GUDN has low sensitivity and is a good candidate for insensitive munitions and gas generators.[18] It is neither soluble in water nor hygroscopic. Its thermal stability is comparable to RDX and is superior to that of ADN. It can also be used in LOVA propellants for artillery modular charges. The calculated performance of GUDN is between TNT and RDX. The good burning characteristics and its high yield of gas make it an excellent gas generator in air bags of automobiles. Only small amounts of oxidizer are needed for the complete burning of GUDN without releasing any toxic carbon monoxide. Studies and testing are in progress for use in insensitive gun propellants.

The use of potassium and cesium dinitramide in combination with titanium powder for special pyrotechnic applications was given by Berger *et al.*[19] The titanium-metal dinitramide salt compositions were investigated for its combustion velocity, heat of formation and safety properties of handling. These compositions are found to be extremely sensitive to impact, thus it can replace the use of heavy metal in primary explosives.

References

1. CJ Fawls, JP Fields, TJ Dunn, *et al.*, Propellant composition comprising nanosised boron particles, *U.S. Patent 6652682 B1*, 2003.
2. (a) CJ Hinshaw, RB Wardle, TK Highsmith, Propellant formulations based on dinitramide salts and energetic binders, *U.S. Patent 5498303*, 1996. (b) CJ Hinshaw, RB Wardle, TK Highsmith, Propellant formulations based on dinitramide salts and energetic binders, *U.S. Patent 5741998*, 1998.
3. M Johansson, John de Flon, A Petterson, *et al.*, Spray prilling of ADN, and testing of ADN based solid propellants, *3rd Int Conf on Green Prop for Space Prop*, 2006.
4. H Edvinsson, A Hahma, H Östmark, FOA report No. FOA-R—00-01636-310-SE, 2000.
5. RP van den Berg, JM Mul, PJM Elands, Monopropellant system, *European Patent 0950648 A1*, 1999.
6. (a) K Anflo, TA Gronland, G Bergman, *et al.*, Towards green propulsion for spacecraft with ADN-based monopropellants, *38th AIAA/ASME/SAE/ASEE Joint Prop Conf and Exhibit*, AIAA-2002-3847. (b) K Anflo, S Persson, P Thormahlen, *et al.*, Green propulsion for spacecraft-towards the first flight of ADN-based propulsion on PRISMA in 2009, *57th Int Astro Cong*, Spain, 2006.
7. (a) F. Morring, Swedish Space Corporation, *Aviation Week & Space Tech*, June 13, 2005, 138. (b) F Morring, Formation Flying, *Aviation Week & Space Tech*, Sep 12, 2005, 38.

8. N Wingborg, A Larsson, M Elfsberg, *et al.*, Characterisation and ignition of ADN-based liquid monopropellants, *41st AIAA/ASME/SAE/ASEE Joint Prop Conf and Exhibit*, AIAA 2005–4468.
9. A Langlet, M Johansson, N Wingborg, *et al.*, Explosive, *WO 98/46545*, 1998.
10. A Langlet, H Ostmark, Melt cast charges, *WO 98/49123*, 1998.
11. G Santhosh, S Venkatachalam, K Krishnan, *et al.*, The phase stabilisation of ammonium nitrate by potassium dinitramide — A differential scanning calorimetric study, *New Trends in Res of Energ Mat, Part 1*, 312–319, 2005.
12. Kim Jun Hyeong, Noh Man Gyun, Seo Tae Seok, Process for preparing phase stabilised ammonium nitrate containing potassium dinitramide, *Korean Patent Appl 1020000015463 A*, 2000.
13. TK Highsmith, CJ Hinshaw, RB Wardle, Phase stabilised ammonium nitrate and method of making same, *U.S. Patent 5292387*, 1994.
14. KO Christe, WW Wilson, Dinitramide anion as a reagent for the controlled replacement of fluorine by oxygen in halogen fluorides, *J Fluorine Chem* **89**: 97–99, 1998.
15. OA Lukyanov, OV Anikin, VA Tartakovsky, Dinitramide and its salts 9. Mercury II dinitramidate, a new reagent in chemistry of organomercury compounds, *Russ Chem Bull* **45**(2): 433–440, 1996.
16. R Haiges, KO Christe, An improved method for product separation in metathetical reactions and its demonstration for the synthesis of anhydrous cesium salts, *Z Anorg Allg Chem* **628**: 1717–1718, 2002.
17. JC Bottaro, MA Petrie, PE Penwell, *et al.*, N,N-dinitramide salts as solubilising agents for biologically active agents, *U.S. Patent 6833478 B2*, 2004.
18. H Ostmark, U Bemm, H Bergman, *et al.*, N-guanylurea dinitramide: A new energetic material with low sensitivity for propellants and explosives applications, *Therm Chim Acta* **384**: 253–259, 2002.
19. BP Berger, J Mathieu, P Folly, Alkali-dinitramide salts part 2: Oxidisers for special pyrotechnic applications, *Prop Expl Pyro* **31**(4): 269–277, 2006.

Chapter 12

PROPELLANT FORMULATIONS BASED ON ADN

Composite propellants based on ADN in combination with energetic binders such as GAP, PBAMO, PNIMMO, PAMMO, PGN and metallic fuels can be realised. The use of ADN in a propellant system would give a dramatic gain in performance over a conventional AP/Al/binder system. Introduction of a more advanced energetic ingredient such as aluminium hydride in place of Al along with ADN would provide a further 12.4% increase in performance. ADN can be put into liquid form and used as a liquid monopropellant which has significantly a higher I_{sp}. This chapter reviews the formulations and applications of ADN based composite and liquid monopropellants.

12.1. Energetic Binders

Binders are polymeric materials with reactive functional groups and are main constituents of solid rocket propellants. They have fairly low glass transition temperatures, good thermal and mechanical properties. In the recent years, many energetic binders and its copolymers were developed for use in propellants. Few of them were attractive in combination with ADN based formulations. The structures of few important energetic binders are shown in Table 12.1.

Energetic copolymers of BAMO/AMMO (50:50), BAMO/THF (60:40), BAMO/NIMMO (73:27) were also realised and are attractive in ADN based

Table 12.1. Energetic Binders Developed in the Past Years

Energetic binders	Structure	Energetic binders	Structure
GAP		PAMMO	
PGN		PBAMO	
		PNIMMO	

formulations. These copolymers have improved mechanical properties and low glass transition temperatures.

12.2. Performance of ADN based Solid Rocket Propellants

Solid propellants based on AP and HTPB usually comprise around 70%–88% AP. Theoretical performance calculations show that the replacement of AP with ADN results in higher specific impulse (I_{sp}). The theoretical maximum I_{sp} for ADN with different energetic binders was calculated using NASA CEAgui thermochemical code[1] and the results are shown in Table 12.2.

It can be clearly seen from Table 12.2, that the use of energetic polymers offered a higher I_{sp} under the experimental conditions used in the study. Higher I_{sp} was achieved for ADN with GAP, PBAMO, PNIMMO and PAMMO at 80% oxidizer level. However, the PGN based propellant showed a higher I_{sp} at low weight % of ADN (~ 66%). In comparison to the AP-HTPB standard reference propellant, I_{sp} increase of 10 sec–13 sec is achieved with energetic binders.

The I_{sp} of ADN with the energetic binders were calculated using the NASA CEAgui thermochemical code for different oxidizer percentages and the results are shown in Fig. 12.1.

Table 12.2. Theoretical Maximum Specific Impulse for ADN Based Propellants

Energetic polymer	Weight (%)	Specific impulse* (sec)
GAP	80.0 (ADN)	306.4
PBAMO	80.0 (ADN)	307.2
PNIMMO	80.0 (ADN)	305.5
PGN	66.7 (ADN)	303.1
PAMMO	85.7 (ADN)	306.4
HTPB**	90.0 (AP)**	294.0

*P_c = 70 bar, expansion ratio = 70
**Reference propellant AP-HTPB

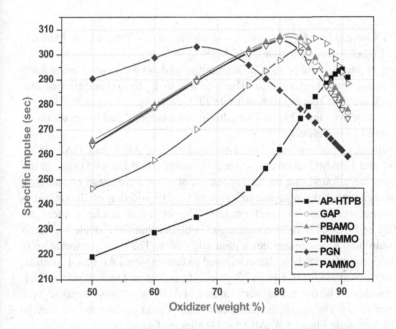

Fig. 12.1. Theoretical performance of ADN based propellants (P_c = 70 bar, expansion ratio = 70).

Figure 12.1 shows the theoretical performance of ADN based propellants with energetic binders in comparison to AP-HTPB propellants. As it can be seen, the I_{sp} of ADN with energetic binders is higher than that of AP- HTPB. Except PGN all other energetic binders showed a maximum I_{sp} at around 80% of ADN. Unlike HTPB which permitted a very high solid loading of up to 90%, all other binders

Table 12.3. Theoretical Performance of Aluminised ADN Propellants

Energetic polymer	Weight (%)	I_{sp}^* (sec)			
		0% Al	5% Al	10% Al	15% Al
GAP	80.0 (ADN)	306.4	315.2	317.3	314.9
PBAMO	80.0 (ADN)	307.2	316.0	318.1	315.3
PNIMMO	80.0 (ADN)	305.5	313.5	315.4	313.9
PGN	66.7 (ADN)	303.1	311.0	315.8	317.2
PAMMO	85.7 (ADN)	306.4	311.4	309.3	—
HTPB**	90.0 (AP)**	294.0	281.3	—	—

* P_c = 70 bar, expansion ratio = 70
** Reference propellant AP-HTPB

except PGN have showed a maximum solid loading of ~80%. However, PGN has showed a higher I_{sp} at low weight % of ADN.

The use of metallic fuels such as aluminium and boron in solid propellants improves the overall performance. The increase in I_{sp} by the addition of aluminium as a metallic fuel is given in Table 12.3.

It can be seen from the table that the addition of a metallic fuel increases the I_{sp} of ADN based formulations.

The theoretical performance calculations involving ADN and GAP, PGN, PNIMMO and PBAMO as energetic binders were carried out by Gadiot *et al.*[2] The authors have found that the incorporation of small percentage of energetic binders shows higher performance in terms of I_{sp}. The addition of aluminium as a fuel further increases the performance. The optimum binder content for ADN/PGN and ADN/PBAMO combination is higher than 20% while for others the optimum binder percentage lies within 10%–20%. The flame temperature of ADN based formulations is higher compared to conventional AP based formulations. At 80% oxidizer level, 14%–17% gain in I_{sp} is obtained with regard to a reference propellant. The theoretical performance of ADN based propellants is given by Schoyer *et al.*[3] The study showed that the propellant combination involving ADN with energetic binder PBAMO gave higher performance.

The theoretical performance of ADN/GAP/Al propellant is given by Bruno D'Andrea *et al.*[4] The authors highlight the perspectives of new generation of solid propellants with enhanced performance eliminating the harmful emission of HCl.

12.3. Composite Propellants Based on ADN

Composite propellant formulations involving ADN and PGN as an energetic binder with aluminium as a reactive fuel is disclosed by Hinshaw *et al.*[5] The authors have found that even at lower percentages of aluminium, ADN based compositions offered high performance with a significant increase in I_{sp}. The authors have compared the performance results of ADN based formulations containing 72% solids where the amount of metallic fuel and the oxidizer is changed at 28% of binder concentration.

Composite propellant formulations involving ADN in the presence of substituted oxirane and oxetane polymers along with reactive metal fuels and other typical propellant ingredients are disclosed by Hinshaw *et al.*[6] The authors have performed a detailed investigation on ADN with PGN in presence of metallic fuels and some propellant ingredients. A comparison of the performance by ADN, AN and AP in a PGN based propellant is also given. In terms of I_{sp}, ADN formulations showed a higher performance. The theoretical performance of ADN and a variety of different energetic binders were also reported by the authors.

Solid propellants based on ADN were investigated by Babuk.[7] The authors have carried out detailed analysis of the performance of ADN, AP, HNF and AN propellants in the presence of binders. A closer study of the specific features of the combustion process for ADN based propellants revealed that real propellants with improved performance can be achieved by ADN based compositions. The effect of ADN particle size, the cost and density of components on the performance of propellants are also given.

Various issues related to the propellant technology of ADN are discussed by Chan *et al.*[8] Propellant formulations of crystals and prills of ADN with GAP, poly(diethyleneglycol-4,8-dinitrazaundeconate) and polycaprolactone polymer containing butanetriol trinitrate (BTTN) and 0.8% n-methyl-p-nitroaniline were studied. The propellants were cured with hexamethylene diisocyanate or a multifunctional isocyanate (N-100) and were aged at 71°C. Their study indicated that ADN based formulations are more prone to deterioration upon aging and the polycaprolactone based propellant appeared to be the best when compared with the rest. All the propellant samples met the shelf life requirement. Using spherical ADN particles, high solid loading can be achieved.

A high energy propellant with reduced pollution using ADN, an energetic binder and a metallic fuel was disclosed.[9] The authors give a detailed study on ADN with energetic plasticisers, energetic binders and mixtures of particulate and ultrafine aluminium. The formulations include GAP, copolymer of BAMO-NIMMO, PNIMMO, PGN, polypropylene glycol and hydroxylterminated polycaprolactone as binders; BTTN, triethylene glycol dinitrate, nitroglycerine and GAP as plasticisers. The use of particulate aluminium in the particle size of 1 µm to 60 µm and ultrafine aluminium in the particle size of less than 1 µm as metallic fuels in the propellant formulations was also explained.

The use of BAMO/NIMMO (70/30) copolymer with ADN is described.[10] ADN in BAMO/NIMMO copolymer with BTTN at 20% aluminium level showed promising performance characteristics than its counterpart oxidizers such as HNF and AP. The authors have compared the I_{sp} values for ADN propellants with BAMO/NIMMO copolymer, PGN, polycyanodifluoramino ethylene (PCDE), bis(nitratomethyl)oxetane (BNMO) and polyethylene glycol (PEG-4000). A comparison of the performance by metallic fuels such as Be, B, Mg, Al with metal hydrides is also made. Based on their study, high energy, minimum-smoke propellants can be made using ADN as a propellant oxidizer.

12.4. Formulations of ADN Based Liquid Monopropellants

Monopropellants are used to maneuver the small thrusters in rocket motors. The extreme toxicity and their limited energy density prompted the search for other alternatives. Advanced monopropellants systems based on ADN have been achieved. The use of ADN for propulsion in small spacecrafts is explored by Gronland et al.[11] Based on numerous studies, a baseline propellant using ADN, water and methanol was evolved. The baseline propellant also contains a stabiliser and gave a I_{sp} of 257 sec, which is higher than hydrazine based monopropellants whose I_{sp} is about 238 sec. The flame temperature of ADN based propellant is nearly twice that of hydrazine based propellants. A novel reactor design has been implemented and tested for ADN based monopropellants. Much experimental verification such as steady state performance, pulse mode performance, blow down capability, reactor preheating and limited life testing were performed on ADN based monopropellants.

A green monopropellant based on ADN is described by Anflo et al.[12] The monopropellants used are mixtures of ADN, a suitable hydrocarbon fuel and a solutant to keep the propellant in liquid state. Green monopropellant candidates based on glycerol, glycine and methanol as fuels with a water content of 26% were studied. The volume specific impulse obtained for ADN/methanol/26% water is 3230 Ns/dm^3. Experimental rocket engine assemblies of 1N-class were designed and tested. The authors have demonstrated that the ADN based monopropellants demonstrate smooth combustion, short ignition delay and short response times. Fundamental function and performance data such as stable combustion, steady state performance, pulse mode performance, system level performance, blow down capability and preheating temperatures were taken into account for the experimental verification.

The formulation of different ADN liquid monopropellants based on ternary ionic solutions and characterisation of their properties such as stability, density, viscosity and sensitivity is reported by Wingborg et al.[13] Monopropellant formulations based on ADN with methanol, and other fuels denoted as F-5 and F-6 were made. The thermal stability, density, viscosity and low temperature properties and sensitivities of these formulations were tested. The results indicate that the formulations based on F-5 and F-6 show superior thermal stability and higher I_{sp}

compared to the monopropellants based on methanol. The two novel formulations are less sensitive towards friction and impact.

The benefits of green propulsion, key requirements and development testing of 1-Newton ADN based rocket engines are described.[14] The firing tests were performed on various ADN based propellant formulations with different equilibrium temperatures. A novel reactor has been fabricated to meet the demands of ADN based monopropellants to achieve a higher combustion temperature (1500°C). The reactor is designed to perform without any significant ignition delay and has functioned for more than an hour during the hot firing tests conducted. Experimental verification of key requirements such as steady state performance, pulsed performance, blow down capability, reactor preheating and limited life testing were carried out.

The characterisation and ignition of monopropellants based on ADN is described.[15] The I_{sp} of FOI liquid propellants FLP-105 and FLP-106 is 10% higher than hydrazine and the density-impulse of FLP-105 is approximately 60% higher than hydrazine. The authors have focused their studies on electrical conductivity, density and specific heat capacity as a function of temperature. The specific heat capacity of FLP-106 is higher compared to FLP-105. Compared to FLP-106 the density of FLP-105 is higher, which is due to the higher content of ADN. The electrical conductivity of these two monopropellants measured between 5°C and 80°C do not follow a linear relation. Samples were analysed using DSC to determine the ignition temperature. The exothermal reaction starts at approximately 150°C. An experimental arrangement has also been made for the electrical ignition study of ADN monopropellant.

A monopropellant system consisting of a formulation containing ADN and water in presence of an alcohol is described.[16] The performance of 50% oxidizer and 50% water mixture can be increased by increasing the amount of dissolved oxidizer and also by adding lower alkanols. The suggested alcohols for the monopropellant systems are methanol or ethanol.

A propellant composition comprising 2%–46% by weight of boron particles having a diameter of 500 nm and a monopropellant comprising the nanosized boron particles to a fuel containing oxygen is reported.[17] The authors have studied the propellant compositions involving the nano-boron particles in presence of few liquid fuels such as ethyl ammonium nitrate, triethylamine nitrate, HMX, TNT and kerosene with oxidising agents such as nitrogen tetroxide, oxygen, hydrogen peroxide, hydroxyl ammonium nitrate, ammonium perchlorate, AN or ADN. By use of the liquid bipropellants based on nano boron-particles, a sufficient gain in I_{sp} and improvement in combustion characteristics of the propellant formulations were achieved.

A new aqueous non-toxic ionic liquid based on ADN is compared with the water oxidizer binary mixtures based on AN.[18] The authors have extensively studied the density of ADN solutions, binary diagram for ADN-water systems, and vapour pressure vs. temperature for the ionic solutions. A ternary system with the addition

Table 12.4. Monopropellants Realised for Use

Monopropellant combination	Additives	Reference
ADN/water/methanol	Stabilisers	[11]
ADN/26% water/methanol	None	[12]
ADN/methanol	F-5 or F-6 fuels	[13]
FOI liquid propellant FLP-105	None	[15]
FOI liquid propellant FLP-106	None	[15]
ADN/water/ethanol or methanol	None	[16]
ADN/liquid fuels	Nano-boron particles	[17]
72% ADN/13.4% methanol/14.6% water	None	[18]

of a fuel to the water-ADN mixture is also studied. The use of methanol as a fuel is described in their studies. The decomposition of binary mixtures of ADN is also studied by thermal analysis techniques. ADN after the complete loss of water showed an exothermic decomposition starting at 142°C with a maximum at 177°C. A small endothermic peak at 225°C is also observed which is contributed to the vapourisation of the formed *in situ* AN during the ADN decomposition. The authors have optimised the formulations of ADN saturated at 20°C. The I_{sp} value obtained is 2698 N.s.Kg^{-1} for 72% ADN//13.4% methanol//14.6% water mixture. The calculated I_{sp} exceeds the value obtained for AN, HAN, HNF and N_2H_4 based formulations. A summary of the monopropellants realised for use is given in Table 12.4.

References

1. S Gordon, BJ McBride, NASA-GLENN chemical equilibrium program, CEA2, 2004, NASA-RP-1311 Part 1, 1994 and NASA RP-1311, Part 2, 1996.
2. GMHJL Gadiot, JM Mul, JJ Meulenbrugge, *et al.,* New solid propellants based on energetic binders and HNF, *Acta Astro* **29**(10/11): 771–779, 1993.
3. HFR Schoyer, AJ Schnorhk, PAOG Korting, *et al.,* High-performance propellants based on hydrazinium nitroformate, *J Prop Power* **11**(4): 856–869, 1995.
4. B D'Andrea, F Lillo, A Faure, *et al.,* A new generation of solid propellants for space launchers, *Acta Astro* **47**(2–9): 103–112, 2000.
5. CJ Hinshaw, RB Wardle, TK Highsmith, Propellant formulations based on dinitramide salts and energetic binders, *U.S. Patent 5498303,* 1996.
6. CJ Hinshaw, RB Wardle, TK Highsmith, Propellant formulations based on dinitramide salts and energetic binders, *U.S. Patent 5741998,* 1998.
7. VA Babuk, VA Vasilyev, DB Molostov, Solid rocket propellants on the basis of ammonium dinitramide: Problems and perspective applications, *33rd Int Annu Conf ICT* **21**: 1–15, 2002.

8. ML Chan, A Turner, L Merwin, ADN Propellant technology, in KK Kuo (ed.), *Challenges in Prop and Comb — 100 Years after Nobel*, Begell House, 627–635, 1997.

9. R Reed Jr, DA Ciaramitaro, High energy propellant with reduced pollution, *U.S. Patent Appln, 2003/0024617*, 2003.

10. ML Chan, Russ Reed Jr, DA Ciaramitaro, Advances in solid propellant formulations, *Prog Astro Aero* **185:** 185–206, 2000.

11. TA Gronland, K Anflo, G Bergman, *et al.*, ADN-based propulsion for spacecraft – key requirements and experimental verification, *AIAA -2004-4145*.

12. K Anflo, TA Gronland, G Bergman, M Johansson, *et al.*, Towards green propulsion for spacecraft with ADN-based monopropellants, *38th AIAA/ASME/SAE/ASEE* joint propulsion conference and exhibit, *AIAA-2002-3847*.

13. N Wingborg, C Eldsater, H Skifs, Formulation and characterisation of ADN-based liquid monopropellants, *Proc 2nd Int Conf on Green Prop for Space Propulsion*, ESA SP-557, 2004.

14. K Anflo, TA Gronland, G Bergman, *et al.*, Development testing of 1-Newton ADN based rocket engines, *Proc 2nd Int Conf on Green Prop For Space Propulsion*, ESA SP-557, 2004.

15. N Wingborg, A Larsson, M Elfsberg, *et al.*, Characterisation and Ignition of ADN-based liquid monopropellants, *41st AIAA/ASME/SAE/ASEE* Joint Prop. Conf. & Exhibit, *AIAA 2005-4468*, 2005.

16. van den Berg Ronald Peter, JM Mul, EPJ Maria, Monopropellant system, *European Patent Appl EP0950648 A1*, 1998.

17. CJ Fawls, JP Fields, TJ Dunn, *et al.*, Propellant composition comprising nanosized boron particles, *U.S. Patent 6652682 B1*, 2003.

18. C Kappenstein, Y Batonneau, E-A Perianu, *et al.*, Non Toxic Ionic liquid as hydrazine substitutes. Comparison of physicochemical properties and evaluation of ADN and HAN, *Proc 2nd Intl Conference on Green Prop For Space Prop*, Italy, ESA SP-557, 2004.

Chapter 13

SOLID COMPOSITE PROPELLANTS AND INGREDIENTS — A COMPARISON

The development and use of solid composite propellants is most important in high energy materials research. With the advent of modern solid propellants in the middle of the 20th century, the maximum performance a solid rocket motor can give has reached a possible value. However, the synthesis of high dense, high performance solid propellant oxidizers and energetic binders in the recent years showed a potential for further increase in energy and performance of solid composite propellants. Much attention is paid to oxidizers because they occupy a large percentage of mass fraction in propellants. Some desirable features of a solid composite propellants include: a high flame temperature, low burning rate exponent, low average molecular weight of combustion products, low temperature sensitivity and a smokeless plume. The performance data and formulation of composite propellants based on ADN with energetic binders and metal fuels were compared and the results are summarised in this chapter.

13.1. Basic Components of a Solid Propellant

A typical solid propellant undergoes rapid and predictable combustion without detonation resulting in large volume of hot gases. The sudden release of these

147

gases through a nozzle is used to propel the rocket engine. For solid propellants, the adiabatic flame temperature during combustion is typically around 2500 K–4000 K. The density is on the order of 1.5 g/cm^3–2 g/cm^3. A propellant must carry enough oxygen for the complete burning of fuel elements i.e, carbon, hydrogen etc. The oxygen is provided by the oxidizer which is essentially oxygen containing compounds such as nitrates, nitramines, perchlorates and nitrate esters. In a solid propellant, the oxidizer particles are bound together by a binder, which acts as a fuel. In addition to oxidizers and fuels, a solid propellant may have other ingredients such as metallic fuels, plasticisers, curing agents, burn rate modifiers etc. A brief overview of the basic components of a solid propellant are given in the following sections.

13.1.1. *Oxidizers*

The primary function of the oxidizer is to provide enough oxygen for the complete combustion of fuel. Oxidizers with low bond energies and large heats of formation are highly desirable. Oxidizers containing N–O, Cl–N or Cl–O bonds are commonly used. The stability of the oxidizer is of great importance during the lifetime and manufacturing of propellant. The density of the oxidizer should be as high as possible to store as much oxygen in a given volume. The basic properties of some solid oxidizers are listed in Table 13.1.

Table 13.1 shows that AP, CL-20 and HMX have high density, the heat of formation is positive for HMX, RDX and CL-20, and is negative for the rest of the oxidizers. HMX, RDX and CL-20 have a negative oxygen balance while others have a positive oxygen balance. The melting temperature also varies, where ADN has a lower melting temperature than other oxidizers. While comparing the heat of formation of ADN and AP, ADN has a more positive heat of formation. The

Table 13.1. Elementary Properties of Solid Oxidizers

Oxidizer	Molecular formula	Oxygen content (wt %)	Density (g/cm^3)	ΔH_f (kJ/mol)	Melting/ decomposition temperature (°C)	Oxygen balance (%)
AN	NH_4NO_3	59.6	1.73	−365.04	170.0	20.0
AP	NH_4ClO_4	54.5	1.95	−296.00	130.0 (decomp.)	34.0
ADN	$NH_4N(NO_2)_2$	51.6	1.80	−150.60	92.0	25.8
HNF	$N_2H_5C(NO_2)_3$	52.5	1.87	−72.00	122.0	13.1
HMX	$C_4H_8N_8O_8$	43.2	1.96	74.88	275.0	−21.6
RDX	$C_3H_6N_6O_6$	43.2	1.82	70.63	204.0	−21.6
CL-20	$C_6H_6N_{12}O_{12}$	43.8	2.04	454.00	>195 (decomp.)	−11.0

densities of ADN and HNF are comparable; HNF has a more positive heat of formation than ADN.

13.1.2. Binders

The role of the binder is to give good mechanical properties for the propellant by means of good adhesion between the oxidizer and metal particles. It is also important that the binder has a lowest glass transition temperature (T_g), because binders with higher T_g would be brittle and difficult to process. The oxygen deficiency of the binder should not be very low. The binder should show no incompatibility with other ingredients used in the formulation of propellants. Some typical properties of binders are given in Table 13.2.

It is seen from Table 13.2 that HTPB and GAP have lowest T_g. GAP, PBAMO and PAMMO have positive heat of formation, PGN and PNIMMO have the lowest oxygen deficiency and PGN has a higher density. The impact sensitivities of these polymers vary, where PNIMMO has a very low value of impact sensitivity. Compatibility studies must be done to select a good oxidizer-binder combination. Polymers having no double bonds in the backbone are best suited for ADN based propellants.

13.1.3. Metal Fuels

Metals contribute to the overall performance of a propellant. The use of metals such as B, Be, Al contributes to the high enthalpy release. The oxidation of Al to aluminium oxide yields over 2 kcal/g. When fluorine containing oxidizers are used, metal fluorides are formed. The enthalpy release of commonly used metals in propellants is listed in Table 13.3.

Table 13.2. Properties of Polymeric Binders

Binder	Formula	ΔH_f (kJ/mol)	T_g (°C)	Oxygen Balance (%)	Density (g/cm³)	Impact sensitivity (Nm)
HTPB	$[C_{10}H_{15.4}O_{0.07}]_n$	−51.9	−63	−323.8	0.92	>50
GAP	$[C_3H_5N_3O]_n$	117.2	−50	−121.2	1.29	16–120
PGN	$[C_3H_5NO_4]_n$	−307.9	−35	−60.5	1.47	>20
PNIMMO	$[C_5H_9NO_4]_n$	−346.4	−25	−114.3	1.26	>9
PBAMO	$[C_5H_8N_6O]_n$	371.5	−39	−123.8	1.30	>20
PAMMO	$[C_5H_9N_3O]_n$	43.9	−35	−170.0	1.06	—

Table 13.3. Enthalpy Release Values of Metal to Metal Oxides and Fluorides

Metal	Oxide	ΔH_f (kcal/g)	Fluoride	ΔH_f (kcal/g)
Be	BeO	−5.8	BeF_2	−5.2
B	B_2O_3	−4.4	BF_3 (g)	−4.0
Mg	MgO	−1.3	MgF_2	−4.3
Al	Al_2O_3	−3.9	AlF_3 (g)	−4.3
H	H_2O (g)	−3.2	HF (g)	−3.2

The effectiveness of metal relies on its high heat release during the oxidation into oxides and it is in the range of Be $\rightarrow$ Al $\rightarrow$ B $\rightarrow$ Mg in a composition. The use of Mg may be effective for oxidizers whose heat of formation is rather low. The use of Be in propellants is limited because of the extreme toxicity of BeO. Introduction of other metals in propellants apart from the ones listed in Table 13.3 always showed a decrease in I_{sp}. It is seen that the use of Be, B and Al results in the highest enthalpy release. Because of the difficulty in burning B and the extreme toxicity of Be, Al is most commonly used as a fuel.

The replacement of metals with metallic hydrides is another alternative to get a higher enthalpy release. The flame temperature is lowered because of the release of evolved hydrogen. The I_{sp} of AlH_3 based propellants is higher by 1 sec–4 sec than that of propellants containing Al. The extreme reactivity and safety hazards associated with the metal hydrides precluded its use in propellants.

13.1.4. *Other Additives*

Plasticisers are added to the binders to improve the processability at low temperatures. The plasticisers must have very low melting point, should be miscible with the binder and if possible contribute oxygen for the combustion. Many of the plasticisers are long chain aliphatic alcohols or acids. Examples include the commonly used diisooctyl adipate (mp −70°C), dioctyl phthalate, diisooctyl sebacate etc. Some of the modern energetic plasticisers butyl-N-(2-nitroxyethyl)nitramine (Bu-NENA), bis(2,2-dinitropropyl)acetal (BDNPA), bis(2,2-dinitropropyl) formal (BDNPF), triethylene glycol dinitrate (TEGDN), trimethylol ethane trinitrate (TMETN), BTTN, 1,5-diazido-3,3-nitroazapentane (DANPE) carry energetic moieties such as $-NO_2$, $-NNO_2$, $-ONO_2$, $-N_3$ etc. which contributes to the overall energy of the composition.

To increase the shelf life of a propellant a suitable stabiliser may be added. These include carbamite, methyl centralite, chalk, diphenylamine etc. A crosslinking agent is also added to facilitate curing of the prepolymer molecule by forming a crosslinked network. It plays a critical role in kinetics of the crosslinking reaction and in achieving the desired mechanical properties of the propellant. Examples are trimethylol propane, glycerol etc.

Apart from crosslinking agents, curing agents such as toluene diisocyanate (TDI), isophoronediisocyanate (IPDI) or hexamethylene diisocyanate (HMDI) are used which react with the terminal functional groups of the binder and the crosslinking agent to give a rigid matrix.

Other liquid or solid additives are often added in small quantities to the propellant composition. Their function is to modify the characteristics of the propellant. For example, burning rate modifiers are used to modify the propellant burning rate and to adjust the pressure exponent 'n' of the burning rate vs pressure curve. Examples are ferrocene, n-butylferrocene, copper chromate etc. Antioxidants are essential to ensure satisfactory aging of the propellant in ambient conditions. Examples are ditertiarybutyl paracresol, 2,2-methylene bis(4-methyl-6-tetriarybutyl phenol) etc. Catalysts are often necessary to reduce the curing time of the propellant. They have significant impact on the mechanical properties by facilitating some favourable reactions, thereby giving direction to the formation of the polymer network. They are usually salts of transition metals. Examples include dibutyltindilaurate, lead octoate, iron acetyl acetonate, lead chromate etc.

13.2. Specific Impulse of Monopropellants

The performance of each solid propellant ingredients have to be thoroughly characterised. It is often necessary to evaluate the performance of the oxidizer as a monopropellant to better understand the combustion and performance of the ingredient by itself. The density of the oxidizer also plays an important role in the overall performance of the energetic formulation. For the oxidizer with a high density, the density impulse would be high, which permits packing more oxidizer in a limited space. Table 13.4 lists the specific and density impulses of most important oxidizers as monopropellants.

Table 13.4. Specific and Density Impulses of Oxidizers as Monopropellants*

Compound	Specific impulse (sec)	Density impulse (sec. g/cm^3)
AP	180.8	352.6
AN	183.0	316.6
ADN	231.7	417.1
RDX	308.1	560.7
HMX	306.5	600.7

*at P_c = 70 bar, expansion ratio = 70.

From Table 13.4, it is seen that in the case of ammonium containing oxidizers AP, AN and ADN, the specific impulse is in the order of ADN > AN > AP; however the density impulse is in the order ADN > AP >AN. It is clear that formulations containing ADN will have a higher performance than AP and AN based formulations. In the case of nitramine oxidizers, HMX has showed a higher density impulse than RDX. The higher specific impulse of these oxidizers is due to the positive heat of formation and high density than that of AN, AP and ADN.

13.3. Formulation and Performance of Composite Propellants

Due to the increase in cost and hazards associated with the formulation of propellants, the performance increase beyond a certain limit is very difficult. Several nitramines with positive heat of formation gave acceptable performance but they increased the sensitivity of the formulations. Energetic polymers and copolymers offered enhanced performance when used with energetic oxidizers such as ADN, HNF and CL-20. AP is moderately energetic because it has a high negative heat of formation (−296 kJ/mol), it evolves HCl during combustion which increase the average molecular weight of exhaust products. AP, which is a workhorse oxidizer for solid propellants, has found enormous application in the field of aerospace. The major problem with the use of AP based propellants is the release of large amounts of HCl during combustion, which is detrimental to the earth's ozone layer and to the atmosphere. Thermochemical calculations of several formulations indicate that ADN based propellant will offer better performance than the conventional AP-HTPB based propellants. Table 13.5 summarises some of the available performance data for few oxidizers with energetic binders along with the data on the emission of HCl.

AP is most widely studied and is the oxidizer of choice for many decades. It is highly stable, has fairly high density and shows no incompatibility problems. AN is very cheap and its heat of formation is low making it as a low effective oxidizer for solid composite propellants. It also shows different phase transitions at the propellant processing temperatures, which is detrimental to the propellant charge. As seen from Table 13.5, composite propellants based on ADN have higher specific impulse than that of AP and AN based formulation. It clearly indicates that the I_{sp} for a particular oxidizer varies according to different formulations. The I_{sp} values shown in the table depend largely on the operating conditions used for the calculation such as chamber pressure and expansion ratio. A good performing composite propellant should contain 18%–20% of metal as a fuel. In order to produce high performing composite propellants, a selection of the right combination of ADN with energetic binders such as GAP, PNIMMO, PBAMO

Table 13.5. Performance Data for Oxidizers with Binders and Metallic Fuels

Oxidizer	Binder (%)	Fuel (%)	Specific impulse (sec)	HCl emission (%)	Reference
ADN	HTPB (15)	Al (20)	274.0	—	[1]
	GAP (25)	Al (20)	276.5	—	[2]
	GAP (20)	—	321.4	—	[3]
	PGN (20)	—	299.8	—	[3]
	PGN (28)	Al (13)	293.5	—	[4]
	PNIMMO (20)	—	317.4	—	[3]
	PBAMO (20)	—	324.0	—	[3]
	PMVT (12)	—	250.0	—	[5]
	HC (15)*	Be (13)	293.0	—	[5]
AP	HTPB (14)	Al (18)	264.5	19.2	[1]
	HTPB (20)	—	276.8	N/A	[3]
	PNIMMO (22)	Al (20)	263.5	17.6	[2]
	GAP (22)	Al (20)	266.0	17.5	[2]
	GAP (22)	Al-Mg (22)	260.3	10.7	[2]
	PGN (28)	Al (13)	282.9	N/A	[4]
	PMVT (18)	—	238.0	N/A	[5]
AN	HTPB (20)	—	246.3	—	[3]
	HC (11)	Al (20)	244.6	—	[6]
	GAP (20)	—	277.2	—	[3]
	PGN (28)	Al (13)	276.1	—	[4]
	PMVT (12)	—	220.0	—	[5]
ARIANE 5 (AP)	HTPB (14)	Al (18)	264.5	21.0	[2]
HMX(10):AP(58)	HTPB (14)	Al (18)	265.9	16.4	[1]
HMX	NG-NC (22)	Al (20)	257.4	—	[6]

*HC — Hydrocarbon binder, N/A — Not available.

etc. has to be made. Because of the low density of ADN than AP, it can be safely used in upper stages of a rocket motor than in the first stage.

It is also seen from Table 13.5 that formulations containing AP release 10%–20% of HCl from each launch. The use of ADN based formulations not only offered improved performance but also eliminated HCl emissions. The development of modern propellants based on ADN maximises the I_{sp} and releases no toxic environmental pollutants.[7] For want of high performance, these propellants can safely be used in modern solid rocket motors.[8]

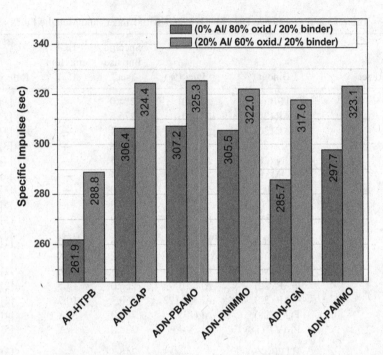

Fig. 13.1. Specific impulse* of ADN with energetic binders in comparison to AP-HTPB (*conditions: $P_c = 70$ bar, expansion ratio = 70).

A comparison of I_{sp} of AP-HTPB and ADN with energetic binders for non-metallised and metallised compositions is shown in Fig. 13.1.

From Fig. 13.1, it is clear that the specific impulse of aluminised propellants is higher than that of non-aluminised. The advantages of using ADN in place of AP can be clearly seen from the chart as the specific impulse for ADN based propellants is much higher than that of AP propellants. The bar chart also indicates that the performance of ADN with energetic binders GAP, PBAMO and PNIMMO are comparable. However, at the binder concentration studied (20%), the ADN propellants based on PGN and PAMMO showed lower performance. The pressure exponent, combustion instability, mechanical properties, sensitivity and thermal stability of ADN based composite propellants are to be studied in detail before being put in use.

References

1. B D'Andrea, F Lillo, Industrial constraints for developing solid propellants with energetic materials, *J Prop Power* **15**(5): 713–718, 1999.

2. B D'Andrea, F Lillo, A Faure, C Perut, A new generation of solid propellants for space launchers, *Acta Astro* **47**(2–9): 103–112, 2000.

3. GMHJL Gadiot, JM Mul, JJ Meulenbrugge, *et al.* New solid propellants based on energetic binders and HNF, *Acta Astro* **29**(10/11): 771–779, 1993.

4. CJ Hinshaw, RB Wardle, TK Highsmith, Propellant formulations based on dinitramide salts and energetic binders, *U.S. Patent 5741998*, 1998.

5. DB Lempert, GN Nechiporenko, GB Manelis, Energetic characteristics of solid composite propellants and ways for energy increasing, *New Trends in Res Ener Mat*, 169–180, 2006.

6. GB Manelis, Possible ways to develop solid propellants for ecological safety, *Hanneng Cailiao (Chinese J Energ Mater)* **3**(2): 9–19, 1995.

7. ZP Pak, Some ways to higher environmental safety of solid rocket propellant application, AIAA-93–1755, 1993.

8. ML Chan, R Reed Jr, DA Ciaramitaro, Advances in solid propellant formulations, in *Solid Propellant Chemistry Combustion, and Motor Interior Ballistics*, *Prog Astro Aero* **185**: 185–206, 2000.

Appendix A

LIST OF ABBREVIATIONS USED

ΔH_f	Heat of formation
2-NDPA	2-nitrodiphenylamine
ABL	Allegheny Ballistics Laboratory
ADN	Ammonium dinitramide
Akardit I	Diphenylurea
Akardit II	Methyldiphenylurea
AN	Ammonium nitrate
AP	Ammonium perchlorate
ARC	Accelerating rate calorimetry
ASTM	American Society for Testing and Materials
BAM	Bundesanstalt für Materialforschung und-prüfung (Federal Institute for Materials Research and Testing)
BDNPA	Bis(2,2-dinitropropyl)acetal
BDNPF	Bis(2,2-dinitropropyl)formal
BHT	Butylated hydroxyl toluene
BNMO	Bis(nitratomethyl)oxetane
bpy	Bipyridyl
BTTN	Butanetriol trinitrate
Bu-NENA	Butyl-N-(2-nitroxyethyl)nitramine
CARS	Coherent anti-Stokes Raman scattering
CID	Collision induced dissociation
CL-20	Hexanitrohexazaisowurtzitane

DABCO	1,4-diazabicyclo[2.2.2]octane
DANPE	1,5-diazido-3,3-nitroazapentane
DBTDL	Dibutyltindilaurate
DDI	Dimeryl diisocyanate
DEA	Dielectric analysis
DF	Density functional
DMF	Dimethylformamide
DMCHA	n,n-dimethylcyclohexylamine
DNA	Deoxyribonucleic Acid
DNEB	Dinitroethylbenzene
DNR	Standard polymeric — MDI
DOA	Dioctyl adipate
DOS	Dioctyl sebacate
DPA	Diphenylamine
DPPH	Diphenyl picryl hydrazyl
DSC	Differential scanning calorimetry
DTA	Differential thermal analysis
EC	Emulsion crystallisation
EGA	Evolved gas analysis
EI	Electron impact
en	Ethylenediamine
ERL	Explosives research laboratory
ESD	Electrostatic discharge
EtOAc	Ethyl acetate
FT-IR	Fourier transform infrared
FT	Fourier transform
GAL	d-Galactose
GAP	Glycidylazide polymer
GUDN	Guanylurea dinitramide
$H_{12}MDI$	4,4′dicyclohexylmethane diisocyanate or Desmodur-W
HAN	Hydroxylammonium nitrate
HC	Hydrocarbon binder
HDI	1,6-hexamethylene diisocyanate
HDN	Dinitramidic acid [$HN(NO_2)_2$]
HMDI	Hexamethylene diisocyanate
HMX	Cyclotetramethylene tetranitramine
HNF	Hydrazinium nitroformate
HNIW	Hexanitrohexazaisowurtzitane (CL-20)
HPLC ·	High performance liquid chromatography
HTPB	Hydroxyl-terminated polybutadiene
Isp	Specific Impulse
IC	Ion chromatography

IPDI	Isophoronediisocyanate
IR	Infrared
KDN	Potassium dinitramide
KF	Potassium fluoride
LAC	Laser assisted combustion
LD	Lethal dose
LIP	Laser induced pyrolysis
LIR	Laser induced regression
LOVA	Low vulnerability ammunition
MBMS	Molecular beam mass spectrometry
MDI	Diphenylmethane diisocyanate
MDSC	Modulated differential scanning calorimetry
MeCN	Acetonitrile
MNA	m-nitroaniline
MOSIAL	Molybdenum supported on silica-alumina
MS	Mass spectroscopy
MT	Microthermocouple
NASA	National Aeronautics and Space Administration
NC	Nitrocellulose
NG	Nitroglycerin
NMR	Nuclear magnetic resonance
NF	No fire
PAMMO	Poly-azidomethylmethyl oxetane
PBAMO	Poly-bisazidomethyl oxetane
PBAN	Polybutadiene acrylonitrile
PCL	Polycaprolactone
PDSC	Pressure differential scanning calorimetry
PEG	Polyethylene glycol
PGN	Polyglycidyl nitrate
PLIF	Planar laser induced fluorescence
PMMA	Polymethylmethacrylate
PMS	Probing mass spectrometry
PMVT	Polymethylvinyltetrazole
PNIMMO	Polynitratomethylmethyl oxetane
PP	Polymeric phosphorous compound
PPG	Polypropylene glycol
PSAN	Phase stabilised ammonium nitrate
PTHF	Polytetrahydrofuran
RDX	Cyclotrimethylene trinitramine
$scCO_2$	Supercritical carbon dioxide
SSC	Swedish space corporation
Tg	Glass transition temperature

TDI	Toluene diisocyanate
TEA	Triethanolamine
TEGDN	Triethyleneglycol dinitrate
TEPAN	Tetraethylenepentamine/acrylonitrile
TFAA	Trifluoroacetic anhydride
TG	Thermogravimetry
TGA	Thermogravimetric analysis
TG-MS	Thermogravimetry-mass spectrometry
THF	Tetrahydrofuran
TIPA	Triisopropanol amine
TISIAL	Titanium supported on silica-alumina
TM	Thermo microscopy
TMD	Theoretical mean density
TMETN	Trimethylethane trinitrate
TMP	Trimethylol propane
TNAZ	Trinitroazetidine
TNB	Trinitro benzene
TNT	Trinitro toluene
TOFMS	Time-of-flight mass spectrometer
TSM	Tripticase soy medium
UN	United nations
UV	Ultraviolet
VC	Verkade's superbase
VIS	Visible
XRD	X-ray diffraction
X1004	Modified monomeric-MDI

Appendix B

LIST OF FIGURES

Appendix C

LIST OF TABLES

INDEX